21 CUSTOM SPEAKER ENCLOSURE PROJECTS YOU CAN BUILD

Other TAB books by the author:

No. 1064 *How To Design, Build, and Test Complete Speaker Systems*

21 CUSTOM SPEAKER ENCLOSURE PROJECTS YOU CAN BUILD

BY DAVID B. WEEMS

TAB BOOKS Inc.
BLUE RIDGE SUMMIT, PA. 17214

FIRST EDITION

THIRD PRINTING

Printed in the United States of America

Library of Congress Cataloging in Publication Data

Weems, David B.
21 custom speaker enclosure projects you can build.

Includes index.
1. Loud-speakers—Amateurs' manuals. I. Title.
TK9968.W44 621.38'028'2 79-25620
ISBN 0-8306-9962-7
ISBN 0-8306-1234-3 pbk.

Contents

Introduction

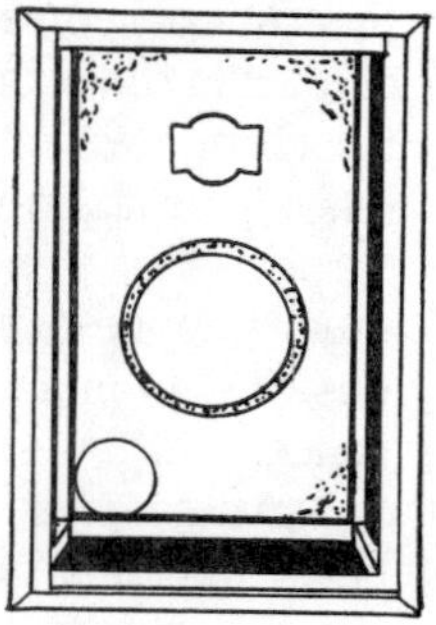

Audio fans often debate about how much money you can save by building your own speaker system. Some manufacturers, wanting complete control over the use of their products, discourage home speaker builders. One reason is they fear that sloppily made enclosures will degrade the sound of their speakers. But there are other, more practical, reasons why certain companies and some audio salesmen try to steer you into a factory-made system. Interestingly, other manufacturers, with equally prestigious names to uphold, gladly furnish information and encouragement to the amateur speaker builder. Fortunately the number of companies in the latter category seems to be growing at a significant rate.

There is no easy answer to the question of how much you can save; much depends on the materials you choose. For example, you can probably buy a good number of factory-made systems in which the figure you pay for the cabinet will be lower than what you must shell out for the enclosure materials of the Omni-Hex. But if you do a careful job on the Omni-Hex, you'll have a fine piece of furniture instead of a mass-produced carbon copy of every other stereo speaker on your block.

Looking through the projects in this book, you will notice that their appearance varies from conventional looking enclosures to some uncommon shapes. In size, these projects vary from mini-enclosures with a cubic volume of one-sixth cubic foot that can fit onto the smallest bookshelf up to an 8 cubic foot floor model. Electrically the spread is just as great, from simple speaker-in-a-box projects to others with crossover networks of varying complexity and up to six or seven drivers.

These projects were chosen because they pleased a panel of listeners. But more than that, listeners were used as guides in making final decisions in project design. For example, in a few projects changing a crossover network component from the theoretically correct value to another value produced better sound, at least in the opinion of several careful listeners. In such a project the change was made. So if you calculate the value of a choke and find it to be different from that specified in the materials list, don't be surprised. If you like to do your own calculations and believe in theory above everything, follow your computer. But remember this, it's an imperfect world—especially where loudspeakers are concerned.

At the beginning of the description of each project you'll find a quick summary of what to expect from that project, both from its difficulty of construction and its sound. Words mean different things to different people, so trying to describe sound character is a tricky business. But if you have definite tastes in the kind of sound you like, the comments for each project should help you choose a speaker system whose performance will please you.

So choose yourself a project and get started. Do a good job, but don't forget to have fun while you're doing it. That's really what this book is all about.

David B. Weems

Enclosures

Engineers sometimes measure the sound of a new speaker while suspending the bare speaker in mid-air. One company even publishes the frequency response curves of such tests. In these curves the response breaks sharply downward below 1000 Hz, showing that a speaker without a box is as helpless as a fish out of water. But when the same speakers are installed in a properly designed enclosure, their response is flat to 100 Hz and within 2 or 3 dB to 50 Hz.

A careful inspection of the two sets of test results show that a box can enable a speaker to deliver 100 times greater sound intensity at 100 Hz than it can produce without the box. Why is a box so important? Does the box produce the extra low frequency power?

Unlike musical instruments, a speaker is a reproducer of sound. It should produce no sound of its own. And a box that makes sound is a bad enclosure. Instead, its function is to block the out of phase sound from the back of the cone so it can't cancel the sound out front. Because low frequency sound waves are naturally omnidirectional, the bass suffers most in an unbaffled speaker. Some enclosures, ported boxes for example, reverse the phase of the backwave and use it to reinforce the bass. But the box itself is silent.

Tweeters usually have an enclosed back, so they need no enclosure except for appearance and proper placement. Many mid-range speakers, too, have a plastic shell over the back of the frame, filled with damping material. If you use an open-back speaker for a mid-range driver and install it in the same box with a *woofer*, you must put a sub-enclosure over the back of the mid-range speaker. Failure to isolate a mid-range driver leaves it at the mercy of air pressure changes that can be great enough to produce increased

Fig. 1-1. Some of the projects described in this book.

distortion and even damage. Each driver needs some kind of enclosure, and the size of the enclosure is related to the cone area of the driver as well as the lowest frequency of sound it must reproduce. Thus tweeters can have flat closed backs, an insignificant cubic volume, but a 15″ woofer requires a box of several cubic feet.

There was a time when almost all speaker enclosures looked alike, but no more. A walk through an audio showroom will reveal an almost endless variety of shapes and sizes of speakers, from the simple to the extreme (Fig. 1-1). Why this diversity? Probably because some manufacturers have recognized that many customers want more than just good sound. In some cases they want a speaker that has great visual interest. Or, like the Pharaoh who built his pyramids upside down, they want something unusual.

DIFFERENT TYPES OF ENCLOSURES

Certain rules should be followed in building a speaker system but almost everyone compromises, to some degree, between optimum audio performance and appearance. For example, a sphere would have some theoretical advantages but not many people own globular speakers. The speaker systems described in this book are designed with respect to proven principles, but they are also designed to be practical.

In Fig. 1-2 you can see several obvious variations in shape possible with a six-sided box. The cube, Fig. 1-2A, is considered one of the poorest choices for a speaker box because it has the same internal distance in three directions. Each of these dimensions would favor the same wavelength, a condition that can produce a

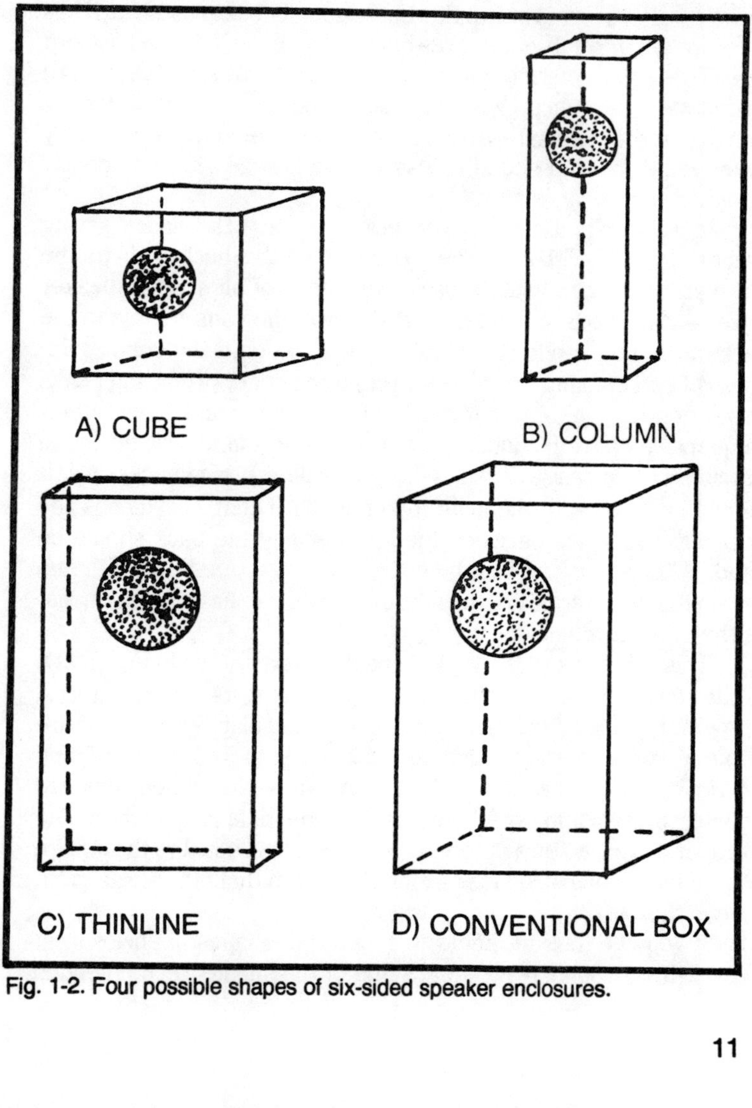

Fig. 1-2. Four possible shapes of six-sided speaker enclosures.

peaky response curve. But later in this book you will see a miniature speaker system with cube-shaped boxes. If the desired cubic volume is very small, a cube may be more desirable than another shape with too little depth behind the speaker.

The second variation, Fig. 1-2B, is prevalent today in column speakers. A simple box with one internal dimension acts like a tuned pipe. That's bad. Such a pipe will produce a frequency response curve with noticeable peaks and valleys. You can stuff the pipe with damping material and modify the roughess, but if you overstuff, you'll kill the bass. Such an enclosure is difficult to debug, so why try? Yes, there is a tower in this book, too. The one used here has been broken into sub-enclosures by a partition, so it doesn't act as a pipe. The tower does have some aural advantages as well as giving a distinctive appearance. Mid-range sound is dispersed more widely from a narrow enclosure than from a wider one, so towers usually have well-dispersed sound unless the mid-range driver is poorly designed or too large.

In some situations it is necessary to use the space saving shallow box, Fig. 1-2C. The nearness of the back wall to the speaker cone aggravates a basic weakness of all simple speaker boxes—reflections. Sound from the rear of the cone travels to the back of the box and is reflected 180 degrees onto the cone again. This reflected sound can travel through the cone, producing peaks at frequencies where it is in phase with the cone and dips at other frequencies. If reflections are serious, the sound will be harsh because of mid-range peaks. Where a shallow box is necessary, it should be filled with damping material. Or if that overdamps the speaker, the space between the speaker and the back should be filled. Where possible, avoid the shallow box. A good rule of thumb is to make the internal depth of the enclousure at least 1½ times that of the speaker itself.

This brings us to the typical speaker box, shown in Fig. 1-2D. While it isn't ideal with its parallel walls and its sharp external corners, it will likely continue as the most popular shape. Here are some dimension ratios that are often used: 5:3:2, 8:5:3 and 14:10:7. Notice that in the first two ratios the dimensions are staggered so that none of them is an even multiple of any other. The third ratio is based on a difference in dimensions equal to the square root of two, which produces a long dimension that is twice as great as the shortest.

If you like to experiment with different shapes of enclosures, Figs. 1-3A through 1-3D show some of the shapes recommended for their acoustical advantages. Several years ago the English speaker authority, G. A. Briggs, rated several shapes on a scale of 10. His ratings of the six-sided boxes shown in Fig. 1-2 were as follows: cube, 3; thinline, 5; and typical box, 7. But he rated hexagonal enclosures at 9 and triangular shapes at a full 10. The reason for the high rating of the triangular shape is that there are no parallel walls and the backwave from the speaker is broken up rather than reflected back to the cone.

Externally the shape is less desirable than that of the rectangular truncated pyramid mounted on a box, Fig. 1-3B. The hexagonal shape offers the same advantages and disadvantages as the rectangular truncated pyramid on a box, good external design but with the opposite wall parallel to the speaker board. That leaves the pentagonal shape which would seem to give the best of both worlds—no internal parallel surfaces and no sharp external corners. The speaker could be installed on one of the sides, or on a sloping top board, as shown. Mr. Alexander Mench, of Fort Worth, Texas, reports good performance from a pentagon. Theoretically, at least,

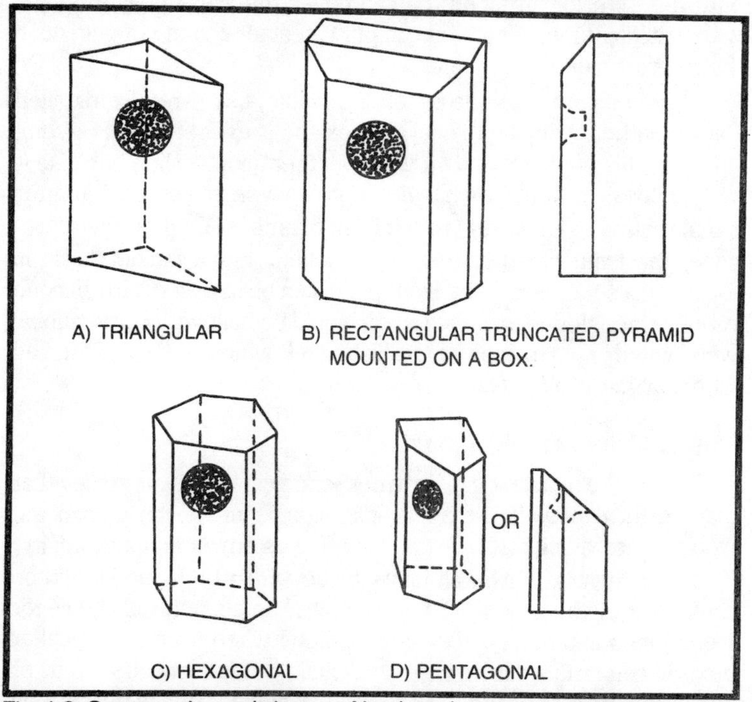

Fig. 1-3. Some rarely used shapes of loudspeaker enclosures that offer good performance.

the pentagon is an attractive choice among the enclosure shapes available to the speaker builder.

CONSTRUCTION RULES

If you decide to build one of the speaker systems in this book but want to change the shape of the box, make sure the internal volume is about the same as that specified in the drawings for the project. You can also make other kinds of changes to fit your tools and materials. Just observe the construction rules that follow.

Rule 1: Make a Rattle-Free Box

The best way to avoid rattles is to do a good job of gluing the joints. Experienced carpenters try to use just enough glue to hold so it won't be smeared onto raw wood and spoil the appearance. That technique will work well enough for most furniture pieces, but it can cause problems in speaker boxes. Unless the wood members are cut to close tolerances, the skimpy use of glue can leave air leaks, not to mention rattles. To avoid the problem of excess glue creeping onto the exterior surface, you can cover the areas near the joint with masking tape. Also, you can put a heavier coating of glue near the interior side of the joint.

If you must make one panel removable, use carefully installed foam weather stripping to seal the box and prevent rattles. Several boxes in this book are shown with weather strip sealing because it was necessary to make changes in crossover or box design after initial testing. The largest reflex is an example. To tune the enclosure, the front panel had to be removed. For a builder, who is following a set of plans, the front panel can be installed with silicone rubber glue. Notice that such a box would be built with conventional wood glued joints except for the last panel, which would be installed with silicone rubber glue.

Rule 2: Make The Box Airtight

Even the smallest gap between joints can cause air noises that mar performance. If the air leak is significant, it can unload the woofer, increasing distortion and limiting its power handling ability. To insure against leaks, all joints made with wood glue should be caulked after the glue sets. Latex caulk is an economical choice for wood joints and has good lasting qualities when used in a speaker box. Silicone rubber is an excellent caulk, but it costs about four or five times as much as the latex caulk. You should choose the silicone rubber to seal speakers to the box. Latex caulk can cause corrosion on bare metal and doesn't adhere very well to painted metal. If you are building a one-piece box, with no removable panels, you can start the box by assembling the sides, top, bottom and the speaker board. Caulk all inner joints before installing the back. Then reach through the speaker holes to caulk the joints around the back.

Apply the caulking material with a gun; then smear it into the joint with a rounded tool. Clean popsicle sticks make good spreading tools.

Rule 3: Choose Quality Material

The most common speaker box material is ¾″ plywood or particle board. Panel thickness should be chosen to match enclosure size. For example, a box with a cubic volume of, say, ¼ of a cubic foot can be built from ½″ plywood or particle board. Such a box will have individual panels small enough that ½″ material will provide adequate stiffness. Unless the panels are unusually narrow, as in a tower, all enclosures for woofers with a diameter of 8″ or greater should be built from ¾″ material. Large panels should be braced for adequate stiffness.

If you use plywood, choose a good grade. Hardwood plywood, such as birch or walnut, is usually well made, but cheap grades of fir plywood will have great voids and poorly glued layers. Voids weaken the panel and loose layers can cause rattles. If you must use fir, ask for "AB" grade of interior plywood. If your plywood has voids, enlarge the enclosure by ½″ each dimension and line the walls with ½″ particle board.

Some of the enclosures in this book show solid pine exterior surfaces. Solid pine is adequate for small boxes or, if lined with particle board, for large enclosures. Unfortunately some of the pine that is sold now as "kiln dried" evidently was rushed through the kiln. This relatively unseasoned wood will warp badly unless you assemble the parts the same day you cut them out. Where you must join more than one board to get the necessary width, as in the largest project, you may find that the boards will shrink as they season, pulling the joints apart. Ideally, such pine should be held in a dry place for several months. If you can find well seasoned wood, pine is an attractive choice for exterior work.

Particle board is especially desirable for good sound. It is dense and has no loose parts or voids. Choose an industrial grade. You can identify the right grade by examining the particle size and the edges of the board. The best particle board for speaker box construction has small particles and smooth edges. Avoid board that is crumbly or flaky.

To some extent your choice of material will dictate your construction technique. For example, if you buy a good grade of ¾″ birch plywood, you will probably want to make beveled joints at the corners to hide the end grain. Here you must be careful in your use of glue to avoid forcing an excessive amount of glue out to the joint and onto the unfinished exterior. Another problem is to clamp the parts together until the glue sets. The simplest way to do that is to wrap something around the box shell to hold the parts together. Adjustable straps are good, but you can use rope, old bicycle tubes, or even tape to apply some degree of pressure to the joints. This kind of joint is likely to be weak, so it's a good idea to add internal glue blocks for reinforcement, especially if the enclosure is larger than 1 cubic foot. Just cut the blocks from ¾″ stock and glue them into each corner, as shown in Fig. 1-4. The blocks can be clamped or screwed into place so that some pressure will be applied while the glue sets.

If you have a power saw, you can make a rabbet joint that will show very little plywood end grain. You can use a set of dado blades to make the cut in a single pass through the saw, but dado blades sometimes cause splintering. A safer method is to use your combination saw blade to remove the amount of material desired. This will require two passes over the saw, one with the material on the table and a second pass with the panel on edge. Always make some practice cuts on waste material to check the setting of your saw blade.

A variation of the rabbet joint is a joint with a solid wood filler at the corner. I call it the "Porter joint" because it was first demonstrated to me by a cabinet maker named Amos Porter. To make it, you set your saw blade to make a cut ½″ deep and set the fence at ½″ from the opposite side of the blade. When you run the panel through the saw twice, first flat and then on edge, you will remove a ½″ × ½″ strip of material. The joining panel is left full thickness. When it sets in the ½″ × ½″ groove, a notch, measuring ¼″ × ¼″, is left at the corner. You can fill the notch with a strip of ¼″ quarter round or cut a piece of ¼″ × ¼″ wood and glue it in the notch. It's a good idea to gently round off the sharp edge to give a smoother corner and to prevent splintering. This kind of joint, shown in Fig. 1-4, can take considerable punishment because of the strip of solid wood that protects the edges of the plywood veneer.

For particle board enclosures the ordinary butt joint is the sensible choice. Use plenty of glue and 6-penny finishing nails. You can sand off any irregularities after the basic box is completed. Then

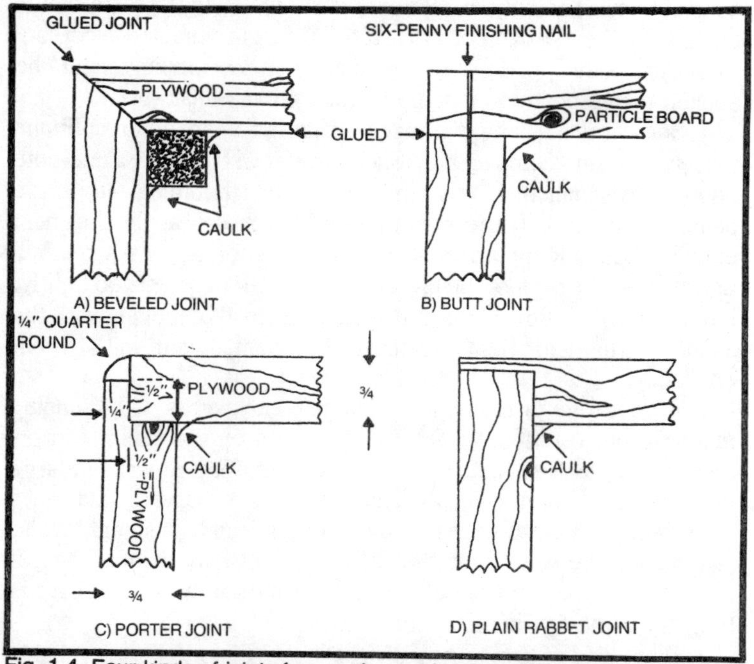

Fig. 1-4. Four kinds of joints for speaker enclosure construction.

cover the box with pine, plastic veneer, real wood veneer, ¼" birch plywood, fabric-or whatever you choose. When you apply a wood covering material, be sure to use enough glue to prevent rattles.

Rule 4: Don't Forget Damping Material

Speakers that don't have enough damping material sound loud. Too much material can overdamp the speaker in some systems, but too little sounds worse, especially in compact enclosures. Some builders use damping material sparingly on the theory that too much kills the high frequency response. Actually it has no effect on the highs because the typical tweeter has a closed back and is isolated from the woofer compartment. Damping material is necessary to absorb sound from the rear of the woofer cone—sound that would otherwise be bounced around the box and reflected out through the cone. When damping material is added to a speaker, it suppresses the mid-range peaks, a change that some listeners interpret incorrectly as a loss of highs.

If you are building a reflex system, you should keep the area around the port opening free of damping material or any other obstructions. The only exception to this rule is any system that is designed to have a damped port. If the reflex is designed according to the data developed by A.N. Thiele, then no damping should be applied to the port and a smooth-sided port is desirable.

Some speakers have a magnet that is too small for optimum performance in a compact box. Such a driver will produce a one-note boom at resonance. If you notice prominent booming, add more damping material. If the boom persists, you can apply the most extreme damping methods, such as a blanket of material stretched tightly over the back of the driver. Make sure it's stretched tightly; it must resist air flow through the fibers instead of moving with the air flow. This treatment may cure the boom, but it will also cut efficiency.

A good rule of thumb is to fill small enclosures with damping material, but use about a 2 to 3″ thickness on the walls of intermediate enclosures—and a 1 to 2″ thickness on the walls of large floor models. For the final adjustment, use your ear. Add more material to any system that sounds loud at modest sound levels. Make sure the walls near the woofer are heavily damped and, if necessary, hang a curtain of material behind the woofer. If bass is thin, try removing some of the material.

What kind of damping material is best? Fiber glass is probably the one material most widely used, probably because its characteristics are well known. There are other materials that are just as good or even better. Long fiber wool is prized by some experimenters because it absorbs sounds at low frequencies better than any other material that has been tested. Unfortunately, it's hard to find and makes a choice home for moths. Polyester fill is an excellent choice, if you need loose material. Felt is good but expensive. Cotton batting has been largely replaced by polyester materials.

Your choice of material may depend on how you are going to use it. For example, loose fibers of glass can damage the voice coil, so if you are filling the box, you should choose polyester material. Speaker manufacturers often frown on the use of house insulation because of the possiblility of loose fibers, but it is probably safe enough when used to line the walls of a large box. One manufacturer suggests covering the back of the woofer with a fine mesh cheese cloth to protect it from loose fiber glass. Another ploy is to install the fiber glass to the walls of the box and then cover it with a sheet of fine mesh cheese cloth. This requires more cheese cloth, but there is no danger that you will change the damping on your woofer by choking the air flow from cone to box. If you choose house-grade insulation, work with it outdoors or with good ventilation. Remove the fiber glass batting from its kraft paper or aluminum foil backing before installing it in a speaker box. House insulation grade fiber glass works well in large enclosures and is much less expensive than special acoustical fiber glass insulation.

For small boxes, use polyester fiber fill, or acoustical fiber glass. You can also get the kind of fiber glass that plumbers sell for installation on furnace heating ducts. It is more dense and less likely to shed fibers than the house ceiling or wall insulation. Remember that the denser the material the less thickness is required, so an inch or so of acoustical or heating duct fiber glass may be the acoustical equivalent of 2 or 3″ of looser material.

Rule 5: Install Drivers From Outside the Box

In earlier days the typical high fidelity speaker was bolted to the rear of the speaker board. In some enclosures that is the only practical way to mount a driver, but if the panel is very thick, it produces a cavity at the front of the speaker. Such cavities can produce a resonance and color the sound. Another problem is the sharp edge of the speaker cutout. Sound waves are diffracted as they move past this square edge, producing interference that makes peaks and dips in the mid-range. Most manufacturers continue to install the speaker gasket at the front of the speaker frame. When asked why, one loudspeaker engineer said that most people can't hear the difference in sound caused by rear mounting. Regardless of that opinion, you should use front mounting when possible. If rear mounting is necessary, round off the front edges of the speaker cutouts.

Another benefit of front mounting is that it permits you to build a more vibration-proof box. When every panel is glued to the others, rather than held by screws, it is almost a guarantee that no rattles can develop with use. Even if screws are tight when the enclosure is finished, the constant bombardment produced by high sound levels can loosen a screwed-on panel.

Some speakers have a mounting gasket on the rear of the frame; many have none there. You can run a bead of silicone rubber around the edge of the speaker cutout and set the speaker on that. The silicone rubber forms a perfect air-tight gasket and even glues the speaker to the board. Before you move it, leave the cabinet on its back for several hours so the glue will set. You don't even have to use bolts or screws, but if T-nuts are supplied with the driver, you can use them. Drill holes in the board at the proper locations and drive the nuts into the rear of the board at each hole. You should use the next size larger drill bit to drill a hole for a T-nut. If you want to install 3/16″ bolts and T-nuts, you would use a ¼″ drill bit. For large woofers, such as one 15″ in diameter, use ¼″ bolts and T-nuts.

In cases where you must install the speakers on a vertical wall, where it isn't practical to lay the box on its back, use pan head sheet metal screws to hold the speakers in place while the silicone rubber glue sets. In all cases wait at least 24 hours after gluing to test the system. Air leaks through the uncured rubber can cause false test results or even cause speaker damage if the speaker is driven to high sound levels. Most speaker/enclosure combinations are designed so that the box loads the driver properly. If that load is removed, the cone can move too far, damaging the cone, the voice coil or both.

Rule 6: Choose an Open Grille Cloth

It is pointless to invest in a good tweeter or super tweeter and then cover it with a blanket. Inspect any grille cloth that you are considering by holding it up to the light. If you can see through it, it is probably a good choice.

A better test is to have someone hang the cloth over your speaker and remove it while you listen. Apply this test with a variety of musical programs. If you can hear any change with the cloth in place, try to find a better material. You can use small pieces of material for this test and cover just the tweeter or the tweeter and mid-range driver. Few materials will affect the woofer's output enough to matter.

GRILLE MATERIALS

Some commercial grille materials are chosen by the speaker manufacturer to modify a speaker's quality. If a mid-range driver has a bad peak, for example, the manufacturer may look for a grille cloth with a weave that absorbs more energy at that frequency than any other. Some small stereo radios with modest speakers make use of this principle. One system that uses 4″ × 6″ speakers has a very smooth sound when the speakers are installed in their factory designed boxes. But in the open the speakers have a nasty peak at 4000 Hz. Close examination of the boxes shows a double slotted front, which is very attractive but hard to see through because of the off-set slots. A rough frequency test showed that the raw speaker had a 10 dB peak at 4000 Hz, the peak extending from a little over 3000 to almost 6000. When the speaker was returned to its box, the peak at 4000 was reduced to a mild 2 dB hump. The slotted front produced some dips at other frequencies, but the overall sound was greatly improved by elimination of the peaks. This is further evidence of an old rule: peaks sound worse to the ear than dips.

While special grille materials are available, you can choose from a wide range of suitable fabrics. One of the best choices for just about any enclosure is to ask for a solid color, brown or black, thin double-knit. Such materials are often used on commercial systems. Decorator burlap is less desirable acoustically, but it's cheaper. In some cases it can, like the slotted cabinet, make an improvement in the sound of your system. But, since almost any material will reduce the high frequency response of your speakers, some people prefer to make an attractive front panel and eliminate the grille cloth. There is a risk here, too. While good transparent grille cloth offers little physical protection for the drivers, it does offer some camouflage for them. Small children often find uncovered cones and domes irresistible. It's a good idea of protection. It won't stop blows, but it will make them seem less available.

FINISHING YOUR SPEAKERS

The final appearance of your cabinets depends on how well you make the parts fit together and, perhaps even more, how much time you spend with sandpaper. Start with a medium grit paper to remove projections and ill-fitting trim or other parts. Then go to fine paper for the overall sanding. To check your work, adjust a bright light at a 30 degree angle to the work and observe the wood from all angles. Small scratches or saw marks that don't show on raw wood seem to leap to the surface after a stain is applied.

One of the easiest finishes to apply is the oiled wood effect. First stain the cabinet to the color that you want; then coat it with a mixture of 3 parts linseed oil and 1 part turpentine. Make up the oil-turpentine mixture; then apply it with a rag soaked with the liquid mix. Allow it to stand for a half hour or more; then wipe it clean with a dry cloth. Sand lightly with a 360 grit wet or dry sandpaper and apply another coat. Do this until you have put three coats on the enclosure. Then rub it down. Repeat the process with a further coat once or twice a year for the first few years.

Another easy finish is paste shoe polish. Stain the enclosure; then add the polish for more color and a wax finish. Brown shoe polish over fruitwood stain gives a reddish, almost mahogany, color. If you want other effects, you should practice with multiple coats of various shades of shoe polish over stains on a piece of waste material that is identical to the wood used in the box.

If you decide to apply the more conventional varnish or lacquer over your stain, you should beware of incompatible combinations. The easiest way to make sure that your stains and final finishes will work together without problems is to buy a stain and varnish of a single brand from a knowledgeable dealer. In choosing varnishes you will have a smooth looking finish without extra care by using a satin or a dull varnish. Such finishes are usually less durable than the same line of varnish in a gloss finish. Some furniture finishers prefer to use gloss varnish for the first two coats. Then they apply a single coat of satin varnish to give the final finish a hand-rubbed effect. Again, try to select the different finishes from the same line to avoid incompatible combinations.

MORE BOX CONSTRUCTION HINTS

If you are making a nailed box, such as an enclosure built from particle board with butt joints, make sure you have a solid surface under your work. Unless your work bench is built like an old time battleship, you can do better nailing by placing the parts on a concrete floor. Even a faint movement in the work surface makes it impossible to get tight-fitting panels. If you are having trouble with bent nails, suspect your support.

It isn't necessary to cut a perfect circle for your speakers, but cutting a true circle is easy if your saber saw has a fence (Fig. 1-5). Drive a small finishing nail into the center point of your circle. Start your saw; then, after getting it in the desired cutting line, stop the saw and install the fence in the saw. Clip the head off the nail and drop the fence over it so the nail protrudes through the hole in the fence. Then tighten the set screw on the saw to lock the fence in place. Start your saw again and it will follow the line around the circle without your guiding it.

Remember that in any enclosure each panel should fit against another panel or a cleat so that there are no gaps. The speaker board and the back should be carefully checked on this. If there are gaps in the cleats, they should be filled with caulking material. In some cases the speaker board and the back will rest on a particle board liner. Just make sure they touch something all the way around their perimeters.

Even if you get all the joints air-tight, your cabinet has one other potential leak, the point where the speaker leads go into the box. If you follow the building plans for the projects shown in this book, you will find that a screw terminal strip is specified. It is installed by drilling two holes, one for each terminal, and bringing

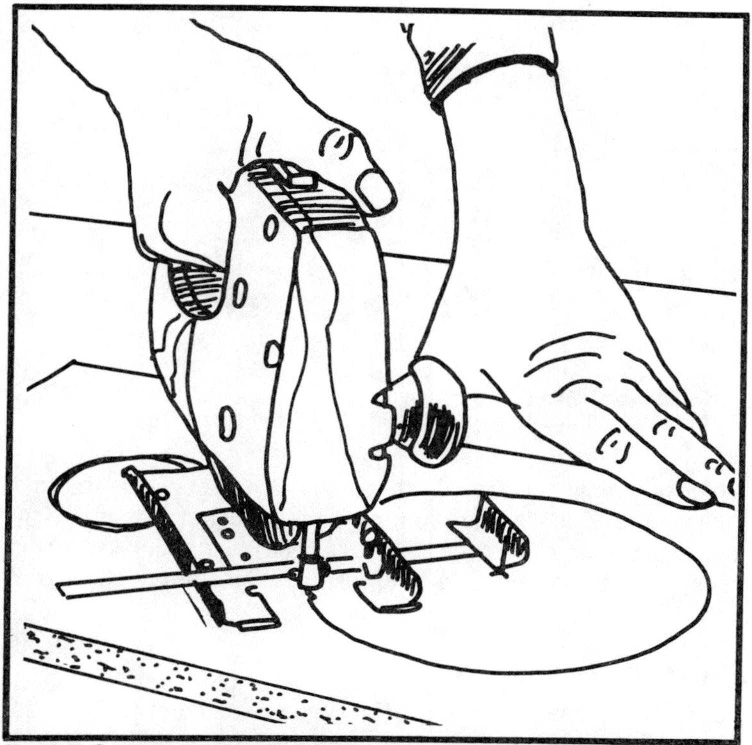

Fig. 1-5. Cutting a perfect circle is easy if your saber saw has a guide.

the speaker leads out through each hole and soldered to the terminals. Then the strip is glued down to the exterior surface of the box. Tacks can be used to hold the strip while the glue is setting. Then check to see that the set screws are all the way in, and fill the terminal holes on the inside of the panel with caulking material. If you have no screw terminal strips, you can drill a single ¼″ hole and pass a lamp cord speaker cable through that. Knot the cable on the inside, to protect the speakers, and fill the hole with caulking material.

When your enclosure is finished, you can check the quality of construction by rapping each panel with your knuckles. You should hear a dead, relatively high-pitched sound. If you hear a drumming, you will know that the panel you are hitting isn't braced well enough. A rattle tells you something is loose. Try this same test on commercial speakers and you will quickly be able to separate the well-made speaker cabinets from the junk.

How To Wire Your Speakers

Careless wiring can degrade the sound of any speaker system. Even worse, it can damage your receiver. The first rule on wiring speakers is to make sure the power is turned off on your receiver or amplifier before you start. Having observed that, here is how to do it right.

KIND OF WIRE

Ordinary lamp cord with #18 gauge conductors is the best choice of wire for most speaker systems. If you have a 4-ohm speaker system, you should get heavy-duty lamp cord, with #16 gauge wire. Actually the precise gauge of cable can be determined only after considering the length of cable you need to reach each speaker and the impedance of the speaker. Lamp cord with #18 gauge conductors is good for distances up to 30′ for 8-ohm speakers, and half that distance for 4-ohm systems. Sixteen gauge wire can be run up to 50′ for 8-ohm speakers and 25′ for those with a 4-ohm impedance. If you keep to these lengths, you can be assured that the speaker cable resistance will be no greater than 5 percent of the total load, a conservative figure. Longer cables than those recommended here can be used. If it is much longer, though, the resistance in the cable can reduce your receiver's ability to damp the speaker system at its bass resonance, increasing boom.

Lamp cord is available in various colors, so you can choose one that will blend with your room walls or floor coverings. If brown or white will do, you can find lamp cord at any hardware or discount store. For other colors you may have to search out a store that stocks supplies for electricians.

If you must run a speaker cable under a rug, you can use flat television antenna wire. Check the gauge of available twin lead and get the one with the heaviest conductor, that is, the lowest gauge number. Avoid the flimsy light-gauge cable that many electronic stores sell as speaker cable.

TERMINALS

Even if you never build a speaker system, you may have to solve the problem of how to connect a system that has non-matching terminals. Most high fidelity receivers and many speaker systems have terminals that hold the cable with adjustable screws or nuts. Some receivers have phono jack speaker terminals. A few European models have jacks that match nothing you ever see in this country. Check the kind of terminals on your receiver before you shop for wiring hardware.

If your receiver has the standard terminals, with an adjustable screw to clamp the conductor, it's a good idea to install spade lugs on your speaker cables. You will find that spade lugs come in various sizes, so you must match lug size to terminal size. If you don't have spade lugs that fit, you can make a "j" hook at the end of the wire. Strip off about 1/2" of insulation from the end of the wire to make the hook. Use a sharp knife and cut gently until the insulation is free. If you angle the knife as if you were sharpening a pencil, you are less likely to cut the wire. To prevent cable break, inspect the bared wire carefully for cut strands. This is especially important for cables inside the speaker cabinet where a failure will require major surgery to repair. Twist the strands together and tin the wire with solder. This will prevent any splayed filaments that can short out your receiver. Form the soldered wire into a hook so it will wrap around the screw without extending toward the adjacent terminal. For speaker systems with push-type quick hookup terminals, you would prepare the cable in the same way but without the hook.

Most raw frame speakers have simple solder terminals. Some of the crossover networks used in this book, notably those produced by Speakerlab, have push-on connectors on each woofer or tweeter cable that require no soldering. When using these, make sure you grasp the connector not the wire. For most speakers you must solder the cables to the speaker terminals.

HOW TO SOLDER

Solder makes a secure connection that won't loosen or develop increased resistance with age. Any kind of soldering iron can be used, but a soldering gun is most convenient. A soldering gun will get hot enough for use in five to 10 seconds, so you don't have to leave it on while you make the physical connections. If you do much soldering, it will probably save enough energy, as compared to an iron, to pay for itself.

Any soldering tool should be kept clean and tinned. If the gun is new, pull the trigger until the tip is hot. Then rub it with a piece of rosin core solder. A dirty tip will conduct the heat poorly and make bad joints. Always clean the tip before and after each use. Keep a piece of sandpaper handy to sand off the metallic oxide that forms on the surface of the tip before you tin the tip. Never let anyone use your electronics gun with acid-core solder. If the tip should become contaminated with acid-core solder, change tips before you use it for electronic wiring of any kind. Acid will corrode soldered joints and impair the electrical connection.

Any rosin-core solder will work for speaker wiring, but try to get one with at least 50 percent tin, such as 50-50 solder. Higher tin ontent, such as 60-40, is even better. Such solder will melt easily and cool easily, making a good joint with less heat.

To solder a connection, make a secure mechanical joint first by threading the wire through the terminal and bending it back sharply against the metal. Heat your iron, or gun, and then hold the tip next to the joint. Stick some solder in the junction of the tip and the joint to improve heat conduction from tip to joint. Then apply the solder to the side of the joint opposite from the iron and hold it there until the solder flows freely through the joint. When you remove the iron, the joint should be smooth and shiny.

PHASING YOUR SPEAKERS

If you hook up a single speaker to a mono amplifier or receiver, it makes no difference which speaker lead goes to each terminal. But when two or more speakers cover the same frequency range in the same room, they must push and pull together or the sound from one will cancel that of the other.

Stereo receivers have coded speaker terminals. You must observe the code at the speakers as well as the receiver end to insure proper phasing. The positive terminal is usually marked with a red dot, sometimes a plus mark. The negative speaker terminal usually carries no marking at all. The negative side of the line at the receiver or amplifier may be marked with a black dot, a "C"for common, or a "gnd" for ground. On some receivers the terminals are marked with numbers a "O" for the common and "4" or "8", for the correct matching impedance, at the positive terminal. Even though these markings are nearly always clearly visible, many stereo systems are connected with the speakers in the two channels out of phase. In fact, you can often find stores that sell audio equipment running their speakers out of phase. Such a connection will produce an annoying sensation in the listener, but the effect is subtle enough that it is difficult to say what is wrong. Some people notice a slight disorientation when listening to out of phase speakers. But the obvious characteristic is a loss of lower bass response.

When long speaker leads are run, there is a greater chance of making a mistake in polarity. Examine the speaker cable to see if there are any identifying marks on the individual leads. Most lamp cord has one wire covered with smooth insulation and one with ribbed insulation. Get in the habit of always using the ribbed lead for the positive side of the line, or vice versa, and you will have no trouble maintaining proper polarity. If the insulation on your speaker cord is unmarked, check each wire for a colored thread in one, or a different wire color.

It is also important to observe polarity in wiring together woofers and tweeters in 2-way systems or the three drivers in 3-way systems to prevent holes in the response curve near the crossover frequency. Unless you use a crossover network that produces a 180 degree phase shift, the various drivers should be connected in phase. If you are using all speakers made by the same manufacturer, the coding will be consistent. A few foreign manufacturers use reverse coding. On woofers or full-range speakers you can check the polarity by connecting a 1.5-volt battery to the speaker terminals. Connect the battery so the speaker cone moves forward when the circuit is completed, and to the rear when it is broken. Then mark the terminal connected to the positive pole of the battery with a red dot or a plus mark.

If, in spite of all your precautions, you find that your speakers sound as if they are working out of phase, try reversing the polarity at one speaker system. Note that you must reverse the wires at one speaker only. If you reverse the polarity on both speakers, their relation to each other will be the same as in the first wiring. Suppose you forget to mark the polarity on the back of the cabinet when you assemble a speaker sytem. What then? Use the battery test and watch the woofer cone movement. If the grille cloth hides the woofer, you can hold a single sheet of newspaper before the woofer and have someone make and break the battery circuit. Or place the two stereo speakers face to face and feed a mono signal with plenty of bass tones to the speakers. Switch the connections on one speaker only. Choose the connection that produces the greatest bass response. Don't forget to turn off your receiver or amplifier while you are making changes in the wiring.

SPEAKER IMPEDANCE

Most speakers have a rated impedance of 8 ohms. *Impedance* is a measure of the speaker's ability to limit current flow; the higher the impedance, the less current will flow through the speaker's voice coil. This tells us that for a given setup, a certain receiver with a certain volume control setting and identical program material, a 4-ohm speaker will permit twice the current flow of an 8-ohm speaker. Most receivers will handle 4-ohm speakers quite well. In fact they will deliver more power to a 4-ohm speaker than an 8-ohm speaker. If the impedance falls below 3 or 4 ohms, the current drain becomes excessive. Make sure the total impedance of your system is greater than about 3 ohms. Just follow this simple rule. Don't hook up more than one 4-ohm speaker system or more than two 8-ohm speaker systems to your receiver. Most receivers have terminals for extra speakers with a switch that permits you to operate either set of speakers or all of them together. In the combined mode the two sets of speakers are in parallel, so the total impedance seen by the receiver will be less than that of either speaker system alone. Note that a parallel wiring hookup provides extra paths for the electrons to flow through. The more paths available, the more current can flow. To see how various speaker hookups can affect total impedance, look at Fig. 2-1. In Fig. 2-1A the two 8-ohm speakers are in parallel, so the net impedance of the two is just half that of one speaker. In Fig. 2-1B the same two speakers are connected in series, which doubles the impedance because every electron that flows in the circuit must travel through the voice coil of both speakers to complete the circuit. A series connection should be used only if the speakers are identical. Deficiencies on one speaker can affect the other, but a series connection is always safer for your receiver than a parallel connection. If you have four speakers, you can wire them in the network shown in Fig.21-C. Here the net impedance is the same as that of a single speaker.

You can check the impedance of unlabeled speakers with an ohmmeter. Find the dc resistance of the voice coil and divide that figure by 0.75. For example, a speaker with a measured dc resistance of 6 ohms would be an 8-ohm speaker. Actually the dc resistance may be somewhat lower than that for some speakers that

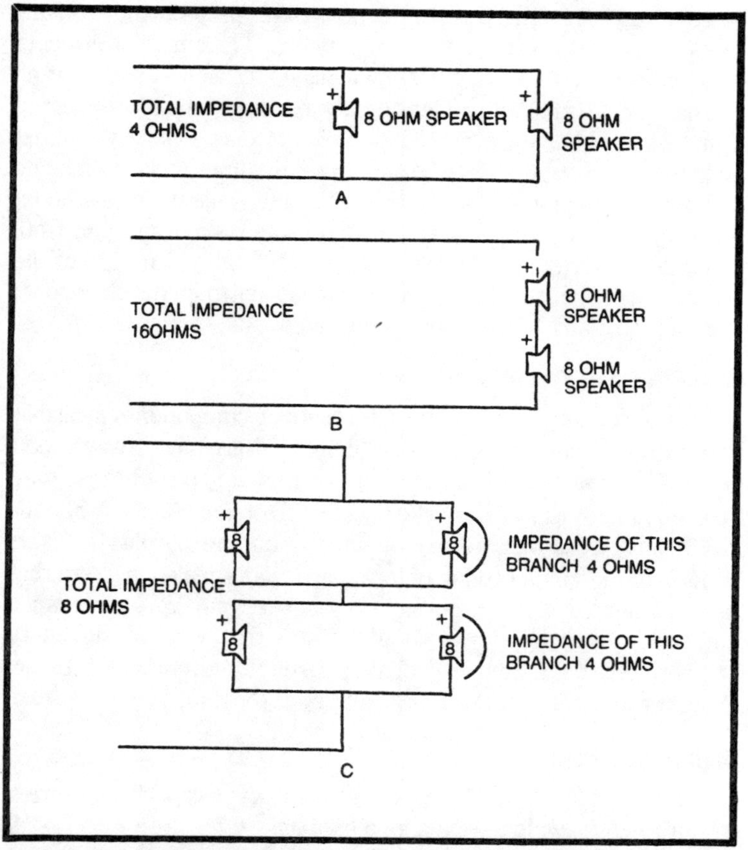

Fig. 2-1. How parallel and series wiring hookups affect impedance. (A). Parallel connection. (B). Series connection. (C). A parallel-series connection.

are rated at 8 ohms. Total impedance consists of pure resistance plus a reactance component that varies with frequency. A speaker that has an impedance of 8 ohms at 200Hz may have an impedance of several times that at its bass resonance, and also at a very high frequency where the inductance of the voice coil produces a high reactance. The important point to consider is the minimum impedance, and for a single speaker that will always be somewhat greater than the dc resistance of the voice coil.

KINDS OF WIRING HOOKUPS

As you look through the projects presented here, you will see some that consist of simple systems, such as a single speaker in a box that requires only two connections. Other projects may require extra wiring but no crossover components. The simplest projects with crossovers use commercially available crossovers so that all you have to do is follow a coded set of wires to get the system wired right. Finally there are more complex systems where you must build the cross over networks, and this sometimes includes making the coils that you use. As you look over the projects in this book, don't let a more complicated wiring hookup discourage you from trying a project that looks interesting. Read the discussion of the project carefully; if you have an average ability to follow directions, you can wire any of the projects in this book.

CROSSOVER NETWORKS

Until recently the number of crossover components available to the speaker builder was severly limited. There have always been generalized commercial crossover networks, and one of these may be adequate for a typical speaker system. But you should be skeptical of any crossover networks offered on the surplus market, especially the kinds produced for a particular speaker system that has now been discontinued. Such crossovers may have impedance correcting or frequency response filters that were designed to complement the particular drivers used by the manufacturer. Whether they will work well with your speakers is anybody's guess.

High-Pass Filters

The simplest kind of crossover network is a capacitor in series with a tweeter which serves as a *high-pass filter.* See Figs. 2-2A through 2-2D. The capacitor offers a high reactance to low frequency energy that could damage small tweeters. At the crossover frequency the reactance of the capacitor is equal to the impedance of the tweeter. Above that frequency the capacitor has a decreasing reactance so that, at very high frequencies, the tweeter is essentially connected directly to the receiver or amplifier. Although some engineers criticize this kind of network as inadequate, others see it as a troublefree method of extending the range of a system without adding inductors to the speaker circuit. Good speaker systems have been built with simple high-pass filters just as other good speaker systems have much more complex networks.

The kind of capacitor used in most crossover networks is the non-polarized *electrolytic*. This kind of capacitor is available in values that range from a few *microfarads* (μf) up to 100 μf or more. Frequently the needed value is not available in a single capacitor, so

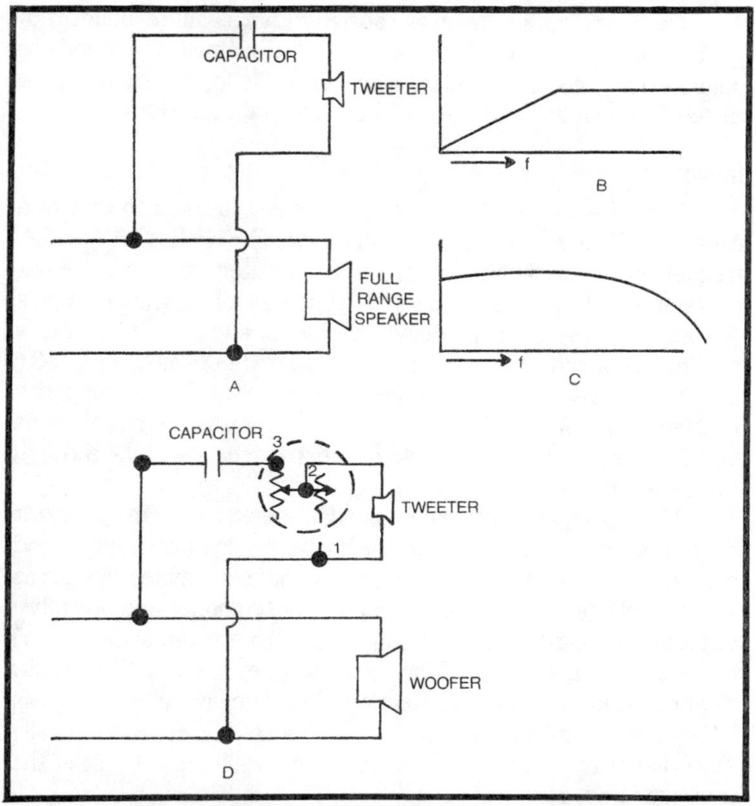

Fig. 2-2. A single high-pass filter is the simplest kind of crossover network. (A). A capacitor acts as a high-pass filter, and a tweeter is added to a full range speaker to extend high frequency range. (B). Tweeter response adds high frequency range to that of full range. (C). A full range speaker has a dropping high frequency response. (D). Adding an L-pad in the tweeter circuit helps to balance the output of the tweeter to that of the woofer.

two or more capacitors may have to be used in paralled. Remember this rule: when capacitors are wired in parallel, their values are added. Suppose you need a 14 μf capacitor, and you find that the only values available are 2, 4, 6, 8 and 12 μf. For the sake of economy you would want to use no more than two capacitors, so you might wire a 6 and an 8 in paralled. You could use a 2 and a 12 or two 6s.

Check the voltage rating on any capacitors that you buy. A capacitor with a voltage rating of 25 working volts would be adequate for a 40-watt system. If you expect to use a more powerful amplifier, get capacitors rated at 50 working volts or greater.

There is another kind of capacitor that is more uniform in quality, the *mylar* type. These are available in small values and are suitable for tweeter circuits. They are also considerably more expensive than are non-polarized electrolytic capacitors.

Inductors

A *coil* has the opposite frequency characteristic to that of a capacitor. Thus a coil in series with the woofer chokes off the high frequencies because its reactance to an alternating current increases with the frequency of the current. One of the simplest kinds of complete crossover networks consists of a coil in series with a woofer and a capacitor in series with the tweeter as in Fig. 2-3.

When a coil is put across the tweeter circuit, it provides a low reactance path for the low frequencies to by-pass the tweeter. At high frequencies the coil's choking action drives the highs through the tweeter.

There are two kinds of coils used in speaker system crossover networks, *air core* and *iron core*. An air core coil may have a solid core as long as the core is made of a non-magnetic material, such as wood or plastic. Air core coils require more copper and are more expensive than iron core coils. Both kinds have advantages. Air coil cores, having more turns of wire, add more resistance to the circuit. That can waste power and affect speaker damping, although it can be good for overdamped speakers. Iron core coils are theoretically more likely to cause distortion, but with well-designed coils the problem is not serious.

Some of the projects in this book specify homemade air core coils; others use commercial crossovers with iron cores. That decision was made primarily by a survey of which kinds were available. You can order ready-made air core coils from the sources listed in the back of the book, or you can make your own.

Band-Pass Filters

When a capacitor is placed in series with an inductor, the combination forms a *band-pass filter*. The capacitor blocks the lows and the inductor chokes the highs. Of course the values of the two components must be calculated so that the filter has the desired bandwidth. When an inductor and a capacitor are combined in a parallel network and wired in parallel to a driver, a similar result is obtained. In that kind of circuit the inductor shunts the lows away from the driver, and the capacitor shunts away the highs. Again, the bandwidth is determined by the specific values of the components.

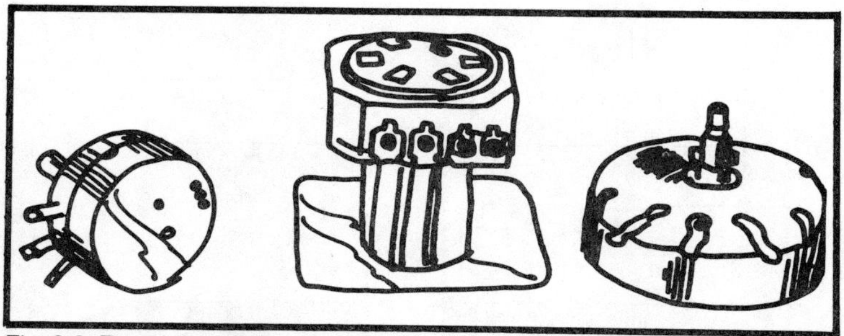

Fig. 2-3. From left to right are L-pads for use with systems of low, medium and high power.

L-Pads

In most 2-way and 3-way systems you will need a control on the mid-range and tweeter to balance them to the efficiency of the woofer. Simple resistors can be used to do this, but require considerable trial and error hookups to get the balance right. Also, they can upset the crossover points by the changing the total impedance of the drivers. An *L-pad* provides an instantly adjustable sound level for the mid-range driver or tweeter, much like a volume control. And it provides a constant impedance to the crossover network and receiver.

Choose your L-pad to match the power rating of your system. If you have a low-powered amplifier and intend to listen at moderate volume levels, the lowest priced L-pads will do, particularly for the tweeter circuit which handles a small fraction of the total power. But if you like your music loud, get a high-powered L-pad, such as one of those pictured in Fig. 2-3. Note that all L-pads serve the same function except that some are suitable for high-powered systems. For that reason you may want to use L-pads with a higher power rating than the ones used in a certain projects in this book. The performance will be identical except at high sound levels.

Use great care in final adjustment of mid-range or tweeter level by L-pad controls. The best way is to turn the level of the drivers all the way down. Then bring up the mid-range level until it balances and integrates with that of the woofer. Do the same with the tweeter. Try to make the system sound as much like a single speaker as possible.

PARALLEL VERSUS SERIES CROSSOVER NETWORKS

Most commercial crossover networks for sale to speaker builders are simple 6 dB per octave parallel networks. In a parallel

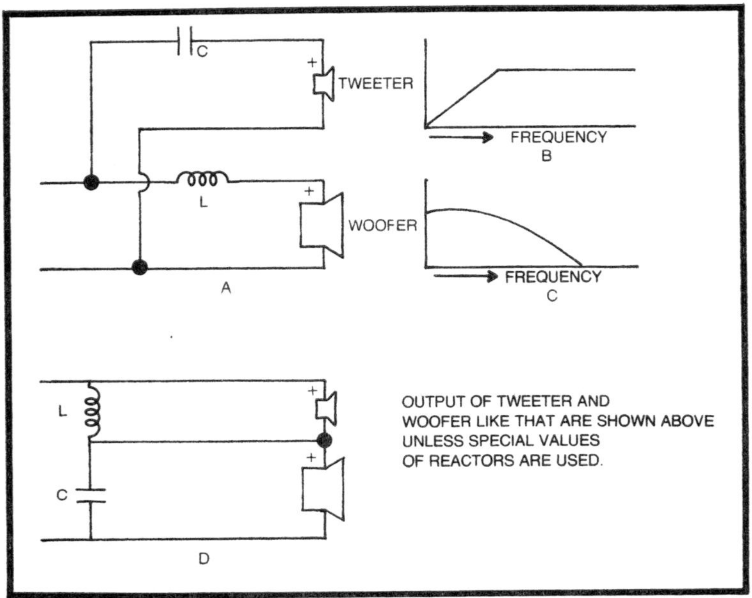

Fig. 2-4. Two-way crossover networks with a single element per channel and 6 dB per octave roll-offs. (A). Parallel circuit. (B). Tweeter output. (C). Woofer output. (D). Series circuit.

network the two speakers appear in parallel with each other. Some people consider series networks to be of doubtful respectability because they have inductors in series with capacitors. At resonance, the detractors say, the impedance across the inductor-capacitor leg will be zero. Nevertheless, many series crossover networks are in use, and regardless of any theoretical problems they seem to work quite well. In fact, like most controversial audio components, they also have their advantages.

Advocates of the parallel network point out that, with each leg of the circuit independent, you can adjust the value of the reactor in one leg without affecting the performance of the driver in the other leg. For example, acoustic suspension woofers typically have long voice coils with considerable inductance. With a parallel network the designer can alter the calculated value of the choke, by subtracting some coils of wire, to compensate for the inductance already in the woofer's voice coil. This may be beneficial in some cases, but it's just as likely that the original value would produce better performance by rolling off the highs more effectively. The final choice should be decided by tests rather than by arithmetic. Partisans of the series networks point out that minor variations in the values of a choke or a capacitor make little difference in their crossovers because the change acts in a complementary way. That is, it shifts the frequency of the crossover point up or down to both woofer and tweeter instead of affecting only one driver. So, although the crossover frequency may vary from the specified value, the circuit will maintain its balance. Another advantage, they mention, is that you can fiddle with the values of a series network and get a somewhat sharp roll-off without adding extra components.

As you look through the projects in this book, you will notice that both kinds of networks are used. For some of the projects several crossover arrangements were tried, and the one that offered the best sound or a good sound with a simpler network was the one chosen. Where the choice of network was considered of importance to the performance of the system, the reasons are reported with the description of the project.

The crossover networks in Figs. 2-4 through 2-6 are simple networks with a roll-off rate of 6 dB per octave. They have the

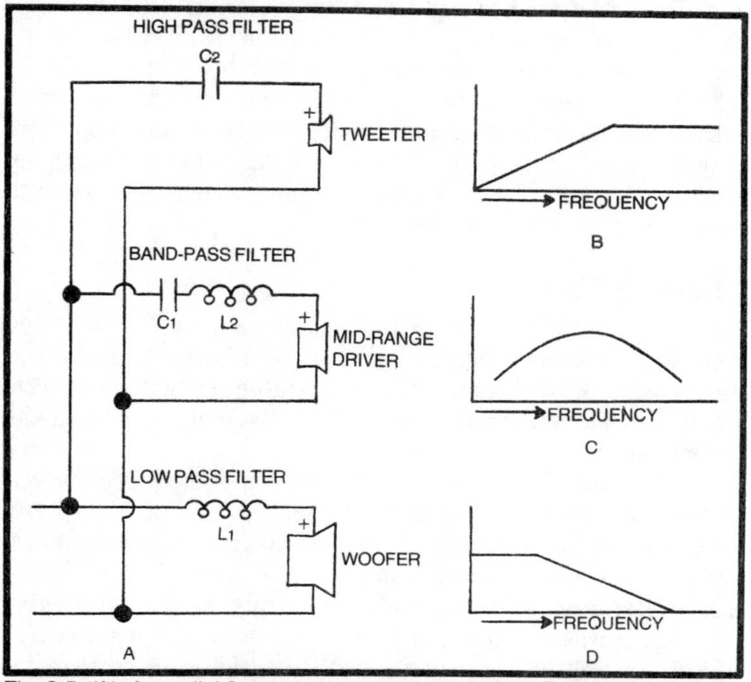

Fig. 2-5. (A). A parallel 3-way crossover network with 6 dB per octave roll-off. Many commercial crossovers available for speaker builders use this kind of circuit. (B). Output of tweeter. (C). Output of mid-range driver. (D). Woofer output.

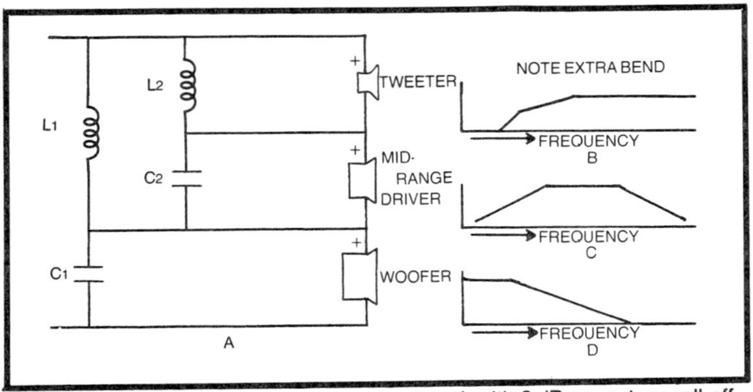

Fig. 2-6. (A). Three-way series crossover network with 6 dB per octave roll-off. Many commercial crossovers available for speaker builders use this kind of circuit (B). Output of tweeter. (C). Output of mid-range driver. (D). Woofer output.

advantage of simplicity, low cost and desirable phase characteristics. Their chief disadvantage is that they permit drivers to perform far beyond the crossover point. This is not necessarily bad unless the drivers have peaks in their response that should be eliminated. The one exception is that some small dome tweeters and other small tweeters can be damaged at high power when connected with a single capacitor to project them. If you substitute a small dome tweeter for another kind of tweeter in any system, note the final section of this chapter carefully.

SPECIAL CIRCUITS

You may notice that some projects have extra crossover components. These are selected to modify some aspect of the particular driver used. If you should happen to substitute another driver, such special circuits would probably be of no value, and it could degrade performance.

One such circuit maintains a level impedance for a driver throughout the audio frequency range (Fig. 2-7A). Figure 2-8 shows the measured impedance of an 8″ woofer before and after an impedance equalizer was added across its terminals. The impedance equalizer is especially useful for woofers and mid-range drivers. It is possible, with an equalizer, to insure that the impedance of the driver will be constant beyond its useful band so that the crossover network will cut off the driver at the desired frequency.

Although most people consider peak suppressors to be useful only for cheap speakers, they can also refine the sound of good quality speakers (Fig. 2-7B). The values of the capacitor and the inductor can be chosen to resonate at the frequency of the speaker's response peak. Such a combination would be all right for sharp spikes, but for most speakers a resistor should be added to load the circuit and broaden its effective bandwidth. Sometimes switching in a peak suppressor can appear to deaden the performance, even if it flattens the response curve. The reason for this apparent degradation is that almost any speaker is at a disadvantage if you lower the volume. Switching in the peak filter will do just that, usually in the mid-range band. To accurately judge the effect of a peak filter one should use some kind of volume compensation, either by simultaneously adjusting the volume control or by switching from the peak filter to a resistive network that equalizes the drive to the speaker.

PROTECT YOUR TWEETER

Any small tweeter, and especially dome tweeters, are subject to damage if you abuse them. The smaller the tweeter the better its high frequency dispersion, but also the more susceptible it will be. Here are some rules you should observe.

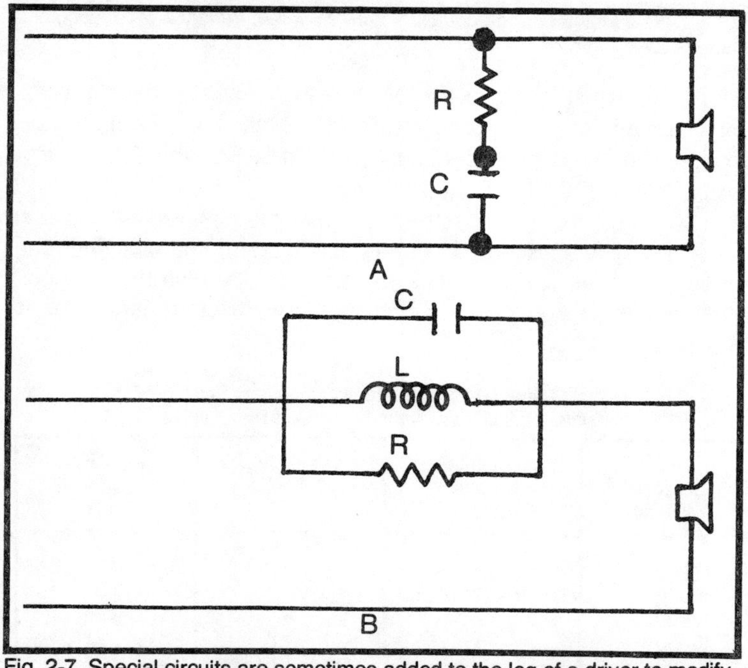

Fig. 2-7. Special circuits are sometimes added to the leg of a driver to modify impedance or response characteristics. (A). Impedance equalizer. (B). Peak suppressor or response shaper, depending on values used.

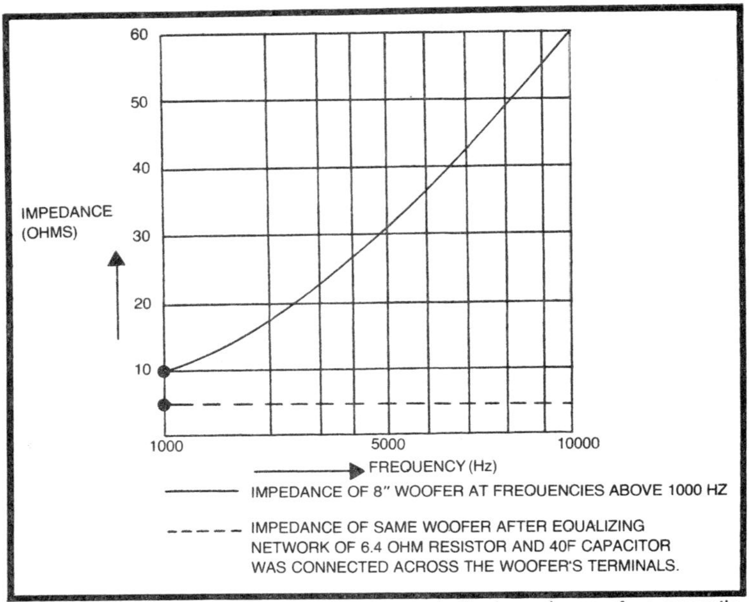

Fig. 2-8. How an impedance equalizer controls the impedance of an acoustic suspension woofer.

■ When you first hook up your speakers, listen to them at very low volume before you turn the controls. Make sure the highs are coming from the tweeter and the lows from the woofer. Almost any wiring mistake can blow the tweeter.

■ Make sure the crossover frequency is high enough for the tweeter. If you substitute tweeters in a system, you may find that the new tweeter is more susceptible to damage than the old one. Small dome tweeters usually require a crossover frequency of at least 5000 Hz.

Table 2-1. Values of new capacitors and added inductor for typical crossover frequencies.

Crossover Frequency	Original Capacitor Value (μf)	Capacitor Value To: (μf)	add L with a value of: (mH)
5000	4	2.8	0.36
6000	3.3	2.3	0.3
7000	2.8	2	0.26
8000	2.5	1.8	0.22
9000	2.2	1.6	0.2
10000	2	1.4	0.18

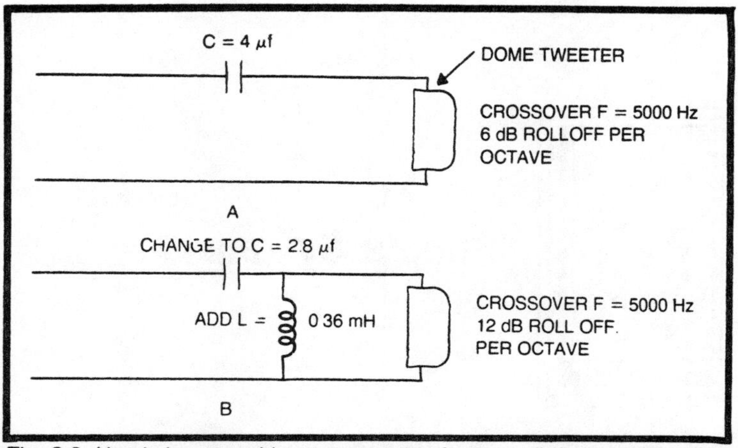

Fig. 2-9. Here's how to add extra protection for a small dome tweeter. (A). Original circuit. (B). Circuit with added protection.

■If you are using a 6 DB per octave crossover network, you can still feed enough low frequency power to the tweeter to damage it. While many tweeters can be used with a 6 dB per octave network, the small dome tweeters need more protection, except in small, relatively low-powered systems. If your crossover has a single capacitor in series with the tweeter, you can change the value of the capacitor and add a choke across the tweeter (Fig. 2-9 and Table 2-1). The new capacitor should be about 0.7 the value of the old one. For example if you have a 4 µf capacitor, replace it with a 2.8 µf capacitor. Then put a 0.36 mH choke across the tweeter. These values would be about right for a 500 Hz crossover.

■If you are building a network yourself, you can use a series network which in a 3-way system gives extra protection for the tweeter (Fig. 2-6). Notice that two chokes are in cascade across the tweeter. Actually the large choke, L_1, appears across both the tweeter and the mid-range driver and the smaller choke, L_2, appears across the tweeter. But, in effect, the two are across the tweeter. This produces an extra degree of low frequency roll-off for the tweeter.

Typical vlaues for 8-ohm speakers and for crossover points at 500 and 5000 Hz would be as follows:

$$L_1 = 2.5 \text{ mH}, L_2 = 0.25 \text{ mH}, C_1 = 40 \mu f, C2 = 4 \mu f.$$

You can experiment with other similar values if such are more available. Put L-pads between the crossover network and the tweeter and mid-range drivers to balance the system.

Simple Systems

The first three projects each consist of a single full-range speaker in a small closed box. This kind of speaker requires minimal wiring, just two connections, and an easy to build enclosure. While bass response is not as deep as in larger systems, these speakers can offer excellent performance for the amount of space they require. Even without the qualification of size, the better simple speakers offer very satisfying music in small rooms. They are especially useful where high power output is not necessary.

CHARACTERISTICS OF THE CLOSED BOX

The closed box is the simplest of enclosures, having no port to require precise alignment. Its most important parameter is the cubic volume of the box. If you get that right, you can vary the dimensions considerably without making a significant change in the sound. You can even make minor changes in box volume without courting disaster. This ability of the closed box to forgive mistakes or variation on speakers makes it a popular choice with speaker builders.

Closed box speakers have been in use since the earliest days of high fidelity, when they were called "infinite baffles" because the total enclosure provided a continuous barrier to the speaker's back wave. The rule then was to build a very large box so that the speaker's resonance would not be moved up in frequency too far by the restricted air volume. In acoustical terms a stiff-coned speaker was matched to a high compliance box. The compliance of a box for a speaker varies directly with the cubic volume of air in the box and inversely with the square of the cone area. So early closed boxes, which had to have high compliance, varied from 1 or 2 cubic feet for small speakers up to 10 or 12 cubic feet for big woofers.

Then, just before the advent of stereo, the acoustic suspension speaker was developed. Instead of putting a stiff speaker in a loose box, the engineers who made the first acoustic suspension speakers reversed the compliance ratios and began putting loosely suspended cones in small or stiff boxes. The air in the box provided the restoring force for the cone that was lost when the suspension was made free. Advocates of acoustic suspension speakers claimed that the air provided a more linear restoring force than any mechanical cone suspension, reducing distortion. Regardless of its acoustic virtues, everyone quickly recognized the advantages of compact boxes particularly when stereo arrived with its demand for at least two speakers. The closed box quickly became the most popular kind of enclosure, and for many years it held that position without serious competition.

In recent years the revival of the ported box has challenged the leadership of the closed box, but sealed speaker systems prevail because of several advantages. As mentioned earlier it is the simplest kind of enclosure to design and build. Where space is critical it can be made extremely small if used with high compliance small speakers. Its bass cutoff rate is more gradual than most other systems, and it doesn't unload the speaker at very low frequencies. Because of the latter virtue, it offers another one. You can drive a closed box speaker from a high-powered amplifier with no worry about subsonic pulses destroying the cone. Finally, it costs less to build. Like all other systems, it does have some disadvantages. Most woofers designed for closed box operation have lower efficiency than some of those used in reflex systems, and the bass range for a given driver is less extended than in a reflex. Considering other sides of the coin, it's no wonder that many speakers are designed for closed box installation.

The three single speaker projects that follow all consist of a small, 4-½" to 6" diameter, speaker in a box that occupies a fraction of a cubic foot. They are all non-critical and should present no problems, even if you have never built a speaker or done any electrical work more complicated than replacing a lamp cord.

PROJECT 1: THE MINI-CUBE

- **Construction:** Easy. Hand tools adequate.
- **Wiring:** Simple. No crossover.
- **Comments:** Pleasant sound in minimum space.

Fig. 3-1. Mini-cubes fit even the smallest bookshelves.

This project is one you can always find room for, a speaker so small it can fit any bookshelf (Fig. 3-1). Although, as mentioned earlier, the cube is considered a poor choice of shapes for large enclosures, here it has the advantage of adding increased depth behind the speaker while permitting reduced box size. This extra depth, when filled with damping material, reduces reflected sound inside the box which can strike the cone and produce dips and peak in the response curve. Regardless of theory, the mini-cube does a good job of providing pleasant sound in locations where you don't have space for a larger speaker.

Choice of Speaker

The speaker specified for this project is a bargain import with a 10 ounce magnet and a high compliance foam suspension. You can substitute other similar speakers of the same diameter. It doesn't matter greatly whether the speaker has a foam suspension, but the suspension must be free. But make sure that the speakers you use have husky magnets and some kind of compliance treatment.

There is one substitution that you should not make—a larger speaker. Even if you could squeeze a 6″ speaker into a mini-cube, you would regret the change. The compliance of an air volume varies inversely with the square of the cone area. With all other factors equal, a speaker with a diameter only 1.41 times that of another speaker will require four times the box volume. So a 6″ speaker would require more than four times the volume for an otherwise identical 4″ speaker. The speaker used here is sold as a 5″ speaker, which is the measurement of the diameter of the frame when measured across the mounting holes. If you measure the frame at the mid-points between mounting holes, the diameter is 4-½″.

To compare the cones of other speakers you should ignore frame measurements and measure the diameter of the effective piston. To get that figure, measure the diameter of the inner cone plus half the outer compliance or surround. An easy way to make the measurement is to start at the inner edge of the gasket on one side of the speaker and include the cone, but not the surround on the opposite side. That measurement on the speaker used in the mini-cube was 3-½″. You can substitute other high compliance speakers if they have an effective cone diameter of 3-½″ or less. If you substitute another speaker, the differences that you can hear will probably be that of minor variations in coloration rather than because of a mismatched box volume.

Construction Procedure

Although large enclosures require ¾″ material, ½″ particle board or plywood is completely satisfactory for the mini-cubes. The enclosure shown was built from ½″ plywood with equal success. See Fig. 3-2 and Table 3-1.

Cut out the parts and make the speaker hole in the speaker board. Paint the speaker board flat black before continuing. Assemble the box with glue and finishing nails. Begin by marking the

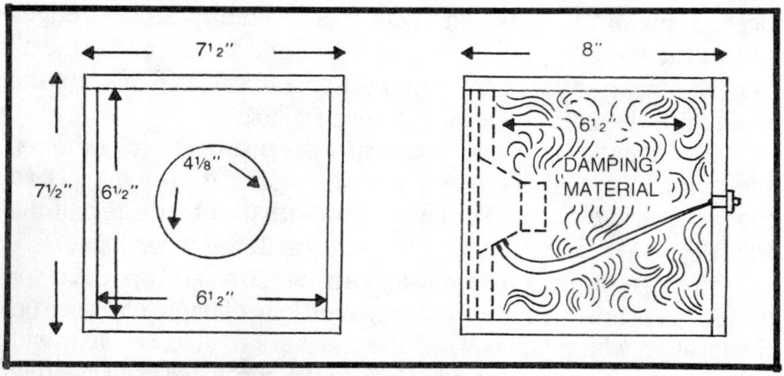

Fig. 3-2. Construction plans for Project 1.

Fig. 3-3. Locate nails on boards before assembly.

position of the speaker board on the inner surface of the sides, leaving a ½″ margin at the front edge for the grille board. Next, draw guidelines on the outside surface of these parts to aid in nail location (Fig. 3-3). These lines should be located ½″ from all edges except the back edge of the top and bottom. Add an extra line 1″ from the front edges of the top, bottom and sides to indicate the position of the inner edge of the speaker board.

Glue and nail one side to the speaker board, keeping the top and bottom edges flush. Next, glue and nail the bottom piece to the side and speaker board. Then follow with the other side and the top (Fig. 3-4). When the glue has set, caulk the inner joints.

Prepare the back by installing a screw terminal strip. Or, if you prefer, measure out a length of lamp cord long enough to reach from the location where you will use the speakers to your receiver with about a foot to spare. Drill a ¼″ hole in the enclosure back and feed

Table 3-1. Materials list for Project 1.

Box material: ½" Particle board. Pieces needed:		
	3	7-½" × 7-½" Back, Top & Bottom
	2	7-½" × 6-½" Sides
	1	6-½" × 6-½" Speaker Board
		Grille board material: ⅜" plywood.
	1	6-7/16" × 6 7/16" Grille board.
Speaker:		5" high compliance. Speaker used here was Westwell 5", Stock #XS-510, McGee Radio Co.

about 10" of lamp cord from the outside to the inside of the back. Tie a knot in the cord to prevent damage to the speaker if the cord is jerked. Fill the space around the cord in the back with latex caulking compound. Install the back on the box with glue and nails. Fill the nail holes with wood putty, and sand the box until it is smooth.

Cut a strip of plastic veneer about ¾" wider than the depth of your enclosure and long enough to reach around it and overlap

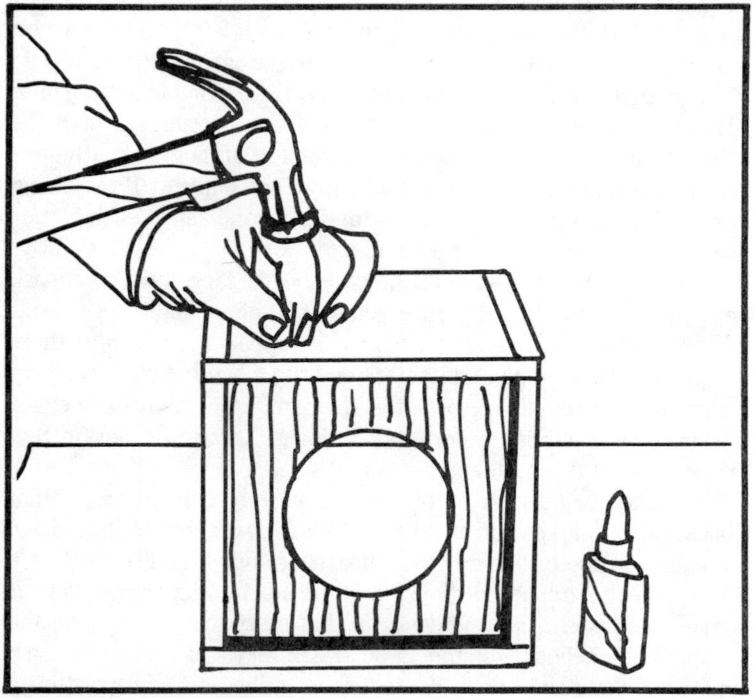

Fig. 3-4. Glue each joint as you nail it.

Fig. 3-5. Soldering the speaker leads to the speaker.

slightly. If your dimensions are the same as those in the plans, this piece of plastic veneer should measure about 8-3/4″ × 31″. Start on the bottom and apply the plastic veneer to the cabinet so one edge is flush with the back edge of the box. Be careful to avoid getting dust or any other foreign material under the plastic veneer. Notch the overhanging veneer at each front corner to an angle slightly less than 45 degrees so that you can fold down one flap and the adjoining one will cover the first. You can trim the second flap at each corner to give the appearance of a bevel joint.

Center the speaker over the grille board, face down, and mark an outline on the board before you install the speaker in the box. Then connect the speaker leads to the speaker and solder them (Fig. 3-5). Set the box on its back and run a bead of latex caulking compound around the edge of the speaker cutout. Set the speaker on the speaker board and press it down. Leave the box in that position until the caulking compound sets.

Make the speaker cutout on the grille board and paint it flat back. When it is dry, stretch the grill cloth over the grille board and staple it on the back side of the board. If your measurements are accurate, the grill board should fit in the box by friction unless your grille cloth is extremely flimsy. In that case you can remove the staples at one edge of the grille board and add a strip of foam material along the edge. Restaple the grille cloth and try the fit again.

When the speaker is ready to use, remove the grille board and temporarily connect a 1.5 volt flashlight battery to the terminals of the speaker. Reverse the connections, if necessary, to find which one makes the cone move outward, or forward, when the connection is made and to the rear when it is broken. Then mark the terminal to which the positive pole of the battery is connected with a plus sign. Do this for each speaker you build.

If you want to use the mini-cube speakers on a bookshelf, you can glue a thin layer of foam on the bottom to protect the shelf and to prevent books from moving the enclosures. Or you can glue a thin, smooth piece of aluminum or stainless steel to the bottom so it will extend under adjoining books. Either of these methods will add enough stability to the mini-cubes to make them into useful bookends.

PROJECT 2: THE OPEN WINDOW

- **Construction**: Use power saw.
- **Wiring**: Simple. No crossover.
- **Comments**: Refined musical sound.

This speaker got the name "open window" because there is no grille or other impediments to block the sound (Fig. 3-6). You can

Fig. 3-6. The finished hardwood speaker board of Project 2 needs no grille cloth.

add a grille if that's what you want, but this little jewel, brushed aluminum set in natural birch, virtually demands to be noticed. After inspecting the handsome little speaker, you will probably agree that a specially constructed cabinet, with finished birch on the top, sides and front, is worth the effort. An added virtue of the box design is that there are no projections at the front to produce diffraction.

CHOOSING BOX VOLUME

The box volume is somewhat smaller than that recommended by Radio Shack, but it was chosen by experimentation. Three enclosures, with cubic volumes of 0.3 cubic foot, 0.5 cubic foot, and 0.8 cubic foot, were built and tested. Listening tests confirmed that the recommended volume, 0.8 cubic foot, gave a performance that was a shade better than that obtained from the 0.5 cubic foot box and noticeably better than that of the 0.3 cubic foot box. But the consensus of opinion among several careful listeners was that the slight difference in performance between the 0.5 cubic foot box and the 0.8 cubic foot boxes wasn't great enough to make the 0.8 cubic foot volume mandatory. The sound from the smallest of the three enclosures was considered less open and a trifle "boxy." And so the middle volume was chosen as a useful compromise between performance and space requirements.

As mentioned above, the appearance of this refined little speaker seemed to cry out for attention. If you want to use less expensive building materials and a different method of construction, that is your option. The plans shown here require a power saw and

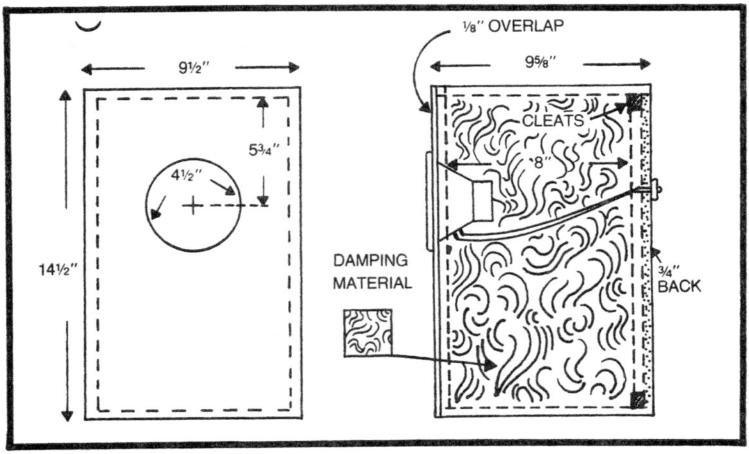

Fig. 3-7. Construction plans for Project 2.

Table 3-2. Materials list for Project 2.

¾" Birch Plywood, Pieces needed.			
	2	9-½" × 14-½"	Sides
	2	9-½" × 9-½"	Top & bottom
	1	9-½" × 14-½"	Speaker board. Dado edges.
¾" Plywood or particle board			
	1	8" × 13"	Back
¾" × ¾" pine cleats			
	2	13"long	Side cleats at back
	2	8" long	Top & bottom cleats at back. (Cut these to fit)
Speaker:	Radio Shack 5" speaker with brushed aluminum frame. Stock number: 40-1311		

some care, but the finished product is unique among small speaker systems (Fig. 3-7 and Table 3-2).

CONSTRUCTION

There are two critical operations in getting a nicely finished enclosure for this project, the beveling of the corner joints and the dadoing of the front board. For these steps you will need a power saw. You can make the dado cut by careful setting of a single power saw blade, but a dado attachment makes quick work of that operation.

Start by setting the fence on your saw at the position that will give you enough boards of 9-½" width to make the top, bottom, sides and front of the enclosure. A single board about 6' long could easily be cut into these pieces. If you are working with shorter pieces you will have to work out how many boards will be required to supply pieces of the right length.

When you have cut enough boards of the correct width, set the saw at 45 degrees to make the end cuts on the top, bottom and side pieces. If you have some waste material that is ¾" thick, it's a good idea to cut some pieces from that and set them on the edge on your work bench to see if they fit together at 45 degrees with no gaps in the joints. Make any minor adjustments in your saw setting before cutting the birch plywood for your cabinets.

Cleats are unnecessary for the front of the box but should be used at the rear. Cut two ¾" × ¾" pine cleats long enough to fit along the flat part of the sides that will lie inside the cabinet. Mark a line inside each cabinet wall, recessed ¾" from the rear edge of the wall. Use glue and nails to attach cleats to the two long sides. An

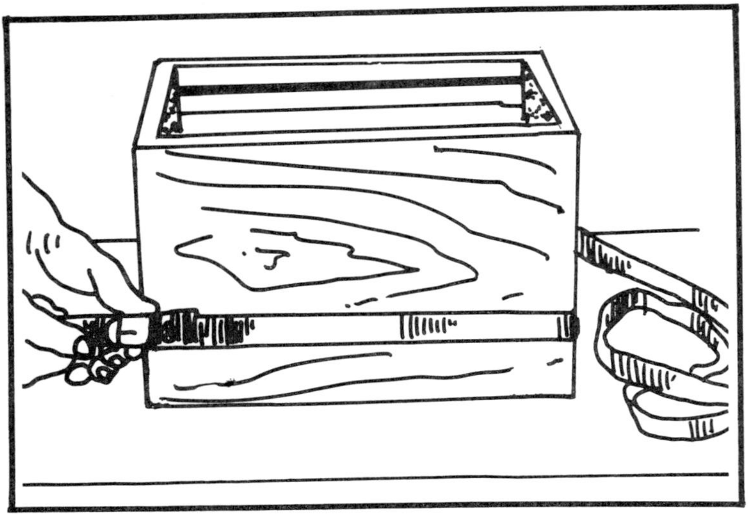

Fig. 3-8. Use straps, ropes or old bicycle tubes to hold the enclosure shell together while the glue sets.

easy way to position the cleats is to lightly nail them in place without glue. Then remove them and apply the glue. If you apply glue before making nail holes in the sides, the cleats will slide around on the slick glue as you drive in the nails. Be sure to use nails that are no more than 1-¼″ long to avoid penetrating the outer surface of the cabinet.

Assembling the Enclosure Shell

After installing the first two cleats, you can apply glue to the beveled edges of the plywood and assemble the enclosure shell. Use straps, ropes or old bicycle tubes to hold the sides together tightly while the glue sets (Fig. 3-8). This method of clamping the parts together works better for bevel joints than ordinary lumber clamps, which will cause shifting of parts as they are tightened.

Check your front board for proper fit on the enclosure and make any minor corrections if it is too large. Make the speaker cutout in the front board. Put a dado attachment on the saw and set it to cut a notch that is ¾″ wide and ⅝″ deep. This will leave a flange on your front board that is ¾″ wide × ⅛″ thick. Test the setting on a piece of waste ¾″ plywood before you proceed. When you are satisfied with the cut, run all four edges of the front board through the dado. Check the board for proper fit (Fig. 3-9). When you have it right, apply glue to the front edge of the plywood walls and to the matching surfaces of the speaker board. Notice that you should apply glue to a strip about ⅝″ wide inside the front edge of the walls and to the ⅝″ deep cut edge of the speaker board, as well as the ¾″ wide strip that overlaps the front edge of the plywood. Use care to prevent the glue from creeping out onto the raw birch veneer.

If you want to make sure that no glue gets onto the birch, you can cover the adjoining surfaces with masking tape. When the board is in place cover it with a piece of plastic or other protective material and weigh it down with some bricks, power transformers or any other heavy objects to keep it under pressure while the glue sets.

Working On the Back

Install a screw terminal strip on the back. Using the terminal strip as a guide, drill two ¼″ holes for the terminal lugs. Prepare a

Fig. 3-9. Speaker board is dadoed to fit into the box with ⅛″ overlap.

piece of lamp cord about 18″ long by removing ½″ of insulation from each end of each lead. Split the conductors apart at one end and pass the leads through the holes in the back. Connect the wires to the lugs of the terminal strip outside the back and solder the connections. Apply a layer of glue to the back around the holes and set the terminal strip on the back. Use small screws or tacks to hold it down while the glue sets. Then caulk the inner holes to prevent air leaks.

Measure the distance between the cleats already installed in the cabinet and cut two cleats to fit in the top and bottom. Attach these cleats with glue, using either screws or clamps to hold them under pressure while the glue sets. Caulk the inner joints of the box, including any gaps between cleats or cleats and box. Fill the box with loose polyester damping material; then feed the lamp cord speaker leads through the damping material to the hole in the front of the box. Glue down the back. If you suspect any gaps, use silicone rubber caulking material to hold the back and seal it. If you use wood glue, apply a liberal coat to the cleats. Set heavy weights on the back to keep the parts under pressure while the glue sets.

Installing the Speaker

Sand down the birch exterior of the box until it is smooth. Apply a stain that matches the appearance you want. For a flawless appearance try a satin varnish or a "dull" varnish. These varnishes give the appearance of hand-rubbed furniture with little effort. Or you can use an oiled finish.

When the cabinet is dry, you can install the speaker. Place the cabinet on its back and solder the leads in the lamp cord to the terminals on the speaker. Then run a bead of silicone rubber sealant around the edge of the speaker hole and set the speaker on the cabinet. Press it down and leave the cabinet in that position for a day. You can trim off the excess silicone rubber.

Connect a flashlight battery to the terminals on the back of the box and observe the speaker cone movement. When you have the battery connected so that the cone moves out (forward) at contact, and to the rear at break, put a plus mark at the terminal to which the positive pole of the battery is connected. Do this for each speaker that you build so you can make sure they are in phase when you connect them to your stereo system.

You'll find the sound of this little speaker is clean and musical. No huge slugs of bass, such as you'd expect from a large woofer, just lucid, refined music that sounds quite natural.

PROJECT 3: THE DUAL CONE SIX

- **Construction**: Use power saw if bevel joints are desired.
- **Wiring**: Simple. No crossover. A frequency shaping filter is optional.
- **Comments**: Good frequency range for size.

This project is built around a high compliance 6″ full-range speaker with a whizzer cone. Two of these systems in stereo give very satisfactory performance. The appearance is more conventional, and certainly less striking than that of Project 2. The cost is lower, too.

Choice of Speaker

The speaker used here is a new version of a model carried in the Radio Shack line for several years (Fig. 3-10). You can identify the current model by noting the letter "C" in the stock number. This

Fig. 3-10. Project 3 enclosure and speaker.

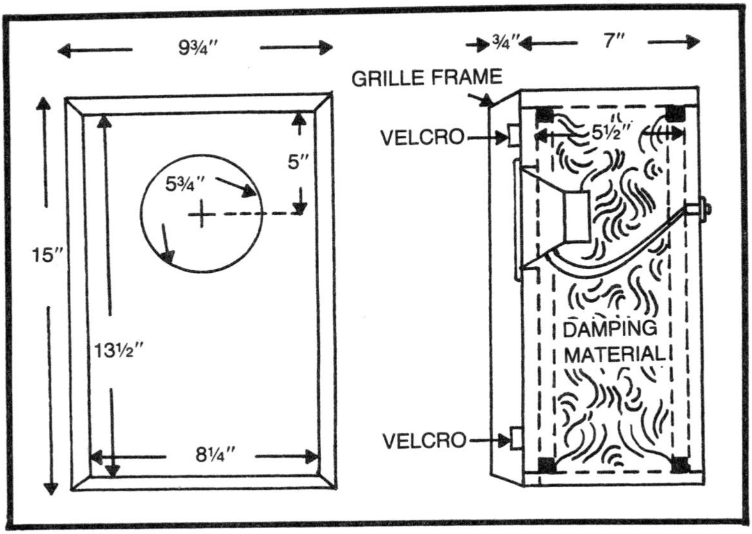

Fig 3-11. Construction plans for Project 3.

speaker has a foam suspension instead of the accordion suspension of earlier models. An old speaker was tested in the box specified here and worked satisfactorily, but the C model is the preferred choice.

Although the earlier speakers had whizzers, the new one had a whizzer that is smaller in diameter than that of the older models. The small whizzer also has a different shape—elliptical instead of round. By installing the speaker with the long dimension of the oval whizzer in the vertical plane, you can obtain improved horizontal dispersion at higher frequencies.

Construction Procedure

Note that you can simplify construction by using butt joints, but in that case you must shorten the length of the sides to maintain the same box dimensions. See Fig. 13-11 and Table 13-3. Cut the sides, top, and bottom pieces to proper width first. Then make the 45 degree bevel cuts at each end of those panels. To insure a true joint, set your saw at 45 degrees and make some practice cuts on waste material. Set two pieces together and check with a square to see if they make a square corner with no gaps in the joint. When you are satisfied that the saw is set right, make the cuts on your box parts.

Cut four 13-½" cleats and attach them to the inside of the side pieces with glue and nails. See Fig. 3-11 and Table 3-3. These cleats

Table 3-3. Materials list for Project 3.

¾″ Birch plywood, pieces needed.		
2	7″ × 15″	Sides
2	7″ × 9-¾″	Top & bottom
¾″ plywood or particle board, pieces needed.		
2	8-¼″ × 13-½″	Speaker board and back.
¾″ × ¾″ pine, pieces needed		
2	9-¾″ long	Grille frame
2	15″ long	Grille frame
4	13-½″ long	Side cleats, front & back.
4	6-¾″ long	Top & bottom cleats, Front & back.
Speaker: 6″ high compliance, Radio Shack #40-1285-C		

should be recessed ¾″ holes through the back to match the position of the lugs on the strip. Run the leads from an 18″ piece of lamp cord through the back, from inside to outside, and solder the leads to the terminal lugs. Install the terminal strip on the outside of the back

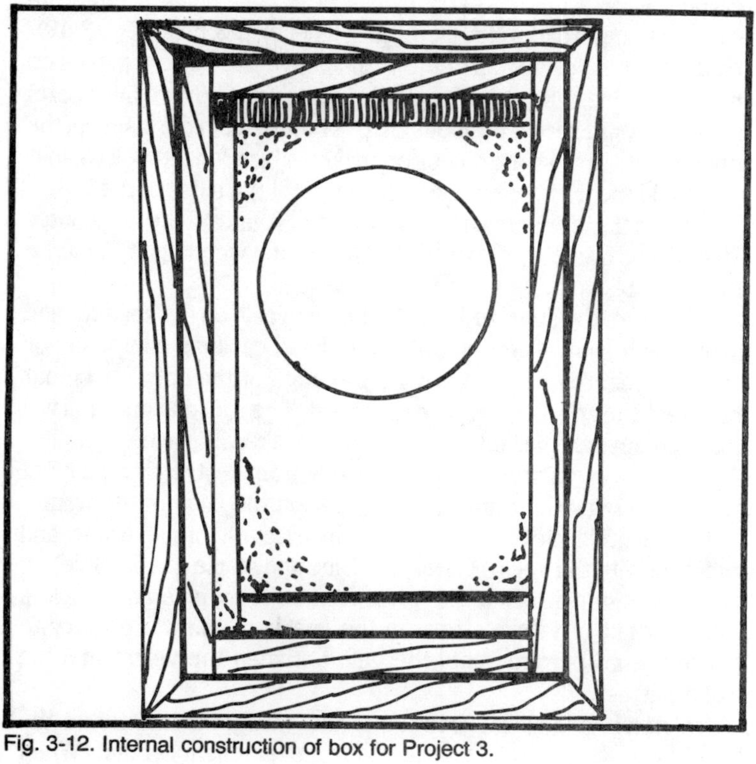

Fig. 3-12. Internal construction of box for Project 3.

with glue and tacks. Caulk the holes on the inside with latex caulking compound.

Cut four more cleats, making the length on each one just right to fit tightly into the space between the long cleats already installed in the box. Glue and screw these cleats into the top and bottom panels. Then install the speaker board with glue and screws.

When the glue has set, caulk the interior joints of the box. Install the back with glue and screws. You can work through the speaker hole to caulk the joints between the back and sides. Then fill the interior of the box with polyester damping material.

Sand the box with a fine grade of sand paper and stain it as desired. Follow this with a wax, oil or varnish finish. Put the box in a dust-free place while the varnish dries (Fig. 3-12).

Grille Frame

Cut four pieces of ¾″ × ¾″ pine to make a grille frame. You can join these pieces with simple butt joints, using glue and small finishing nails. Cut two extra pieces of ⅝″×¾″ pine to hold the velcro patches that will attach the frame to the box (Fig. 3-13). Install these pieces under the top and above the bottom pieces on the grille frame. The ⅝″ dimension permits you to place these parts so that they are recessed about 1/16″ or more from the inside of the grille frame. This slight recession makes space for the velcro so it won't hold the frame away from the edges of the sides of the box.

When the glue sets on the grille frame, round off the front outer edges of the frame with coarse sand paper or a wood rasp. Paint the frame flat black.

Connect the speaker leads to the terminals on the speaker and solder each joint. Then run a bead of latex caulking compound or silicone rubber compound around the edge of the speaker cutout and press the speaker down on the board. The caulking material will make an air-tight gasket and hold the speaker in place.

When the paint is dry on the grille frame, stretch a piece of grille cloth over the frame and staple it on the back of the frame. Trim the grille cloth so it doesn't overlap the grille frame and interfere with the velcro fasteners. Glue and staple a strip of velcro on the blocks just inside the grille frame. Then glue and staple a matching set of velcro strips on the speaker board. Don't try to attach the grille frame until the glue between the velcro and the wood has set.

Connect a flashlight battery to the speaker terminals on the back of each box and observe the cone movement at contact. When

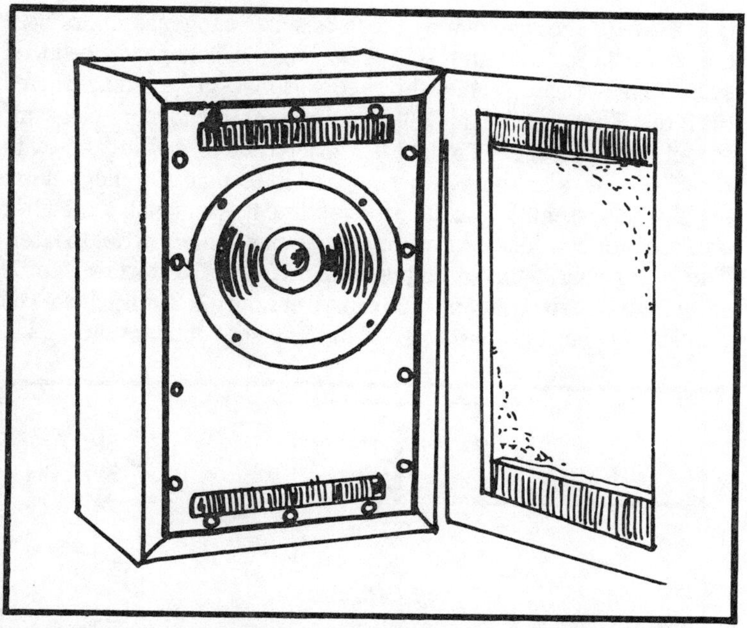

Fig 3-13. Use strips of velcro, installed as shown, to retain grille frame.

the battery is connected so the speaker cone surges forward at contact, and to the rear when the circuit is broken, mark the terminal to which the positive pole of the battery is connected with a plus sign.

Install the grille cloth on your speakers and connect them to your receiver. If you did a reasonably careful job of making the boxes, you will probably be delighted with the sound of these low-cost speakers.

Optional Filter

Most full-range speakers, including expensive ones, produce more sound in the upper mid-range than some people like. The ear is especially sensitive to those frequencies, so even though the speaker has a smooth response curve, many listeners appreciate some degree of cut in the upper mid-range. Too many people accomplish this cut by use of the treble control. If the problem is in the mid-range, this only dulls the performance by rolling off the upper highs until they are lost.

Although the sound of the first Dual Cone Six was surprisingly smooth, an upper mid-range filter was added to one speaker

(Fig. 3-14A). The results were inconclusive. Younger listeners liked the sound better without the filter, saying the filter gave a reduced sense of presence. People who liked soft, background music loved the filter. They said it made the sound smoother and more pleasant. In fact, some listeners thought the string tone of this speaker with the filter was warm and live, comparable to that of a much more expensive system. If you like to experiment, you can try the filter on a speaker and decide for yourself whether to filter or not to filter. The filter requires three components: a 15-ohm 10-watt resistor; a 4.7 μf non-polarized electrolytic capacitor, and a 1.25 mH choke coil (Table 3-4). The only part that will be difficult to obtain is the choke.

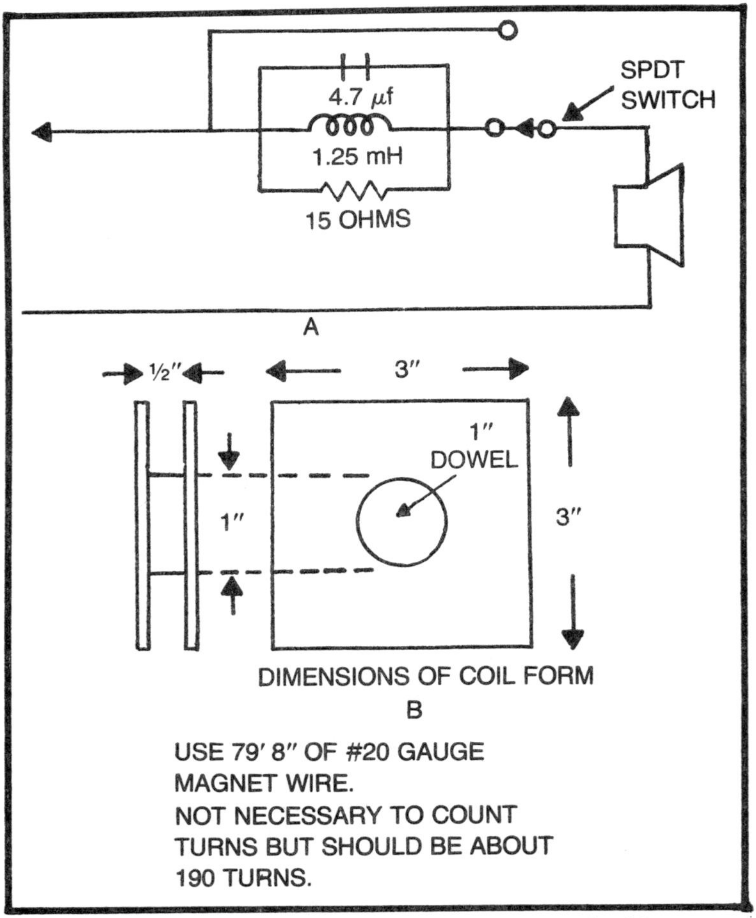

Fig. 3-14. (A). Optional filter for Project 3. (B). Dimensions of coil form.

Table 3-4. Components needed for the filter.

1	4.7 μf non polarized electrolytic capacitor (Radio Shack 272-998 or equiv.)
1	15 ohm 10 watt resistor
1	1.25 mH choke
1	SPDT switch, wired as above to eliminate filter when it isn't wanted.

You can order chokes from some sources, listed in the appendix of this book, or you can make your own. The exact inductance of the choke isn't critical, so making a choke is easy. For two speakers you will need exactly ½ pound of #20 magnet wire, or 160′. To wind a coil, make a form to the dimensions shown in Fig. 3-14B. You can use a piece of broomstick, or 1″ dowel for the core with pieces of thin plywood or hardboard glued to it. Drill a small entry hole through one side of the coil form near the dowel. Measure out 80′ of wire and feed one end from inside the coil form through the hole to the outside. Begin winding, keeping the turns taut. You don't have to count the turns, but the total number should be about 190. When you have 2″ of wire left, tape the coil.

Note that the coil, the capacitor and the resistor are in parallel with each other. The network is in series with the speaker. When you have assembled the filter, you can temporarily connect it in one of the speaker leads to test it. Because it reduces the sound level slightly, you should adjust the volume control upward when the filter is in the circuit to obtain the same loudness level. Give it a fair test; then if you want to use it, you can install it inside the speaker box. Even if you have glued all parts on the box, you can pry up the speaker and install the filter in one lead. You should attach the filter components to the back or bottom of the box with Elmer's Heavy Grip Cement or Leech's Heavy Adhesive. Then replace the speaker with silicone rubber sealer. This optional filter is the final step in your speaker system.

2-Way Closed Box Speakers

Most woofers designed for closed box operation have poor high frequency response. Only the very small high-compliance speakers can be considered full-range speakers, although a few 8″ models do meet that goal fairly well. The heavier cones used in modern woofers do not respond to high frequency signals, and the longer voice coils of the acoustic suspension woofers add a considerable amount of inductance to the electrical circuit, killing the highs. Even if a large speaker can reproduce *treble*, its dispersion characteristics will be poor. Given all of these problems, it's no wonder that nearly all speaker systems with a woofer of 8″ or greater in diameter have tweeters to reproduce the highs.

CHARACTERISTICS OF 2-WAY SYSTEMS

The simplest multi-speaker combinations are 2-way systems consisting of a woofer and a tweeter. Some manufacturers argue that 2-way systems can give performance that is equal to that of 3- or 4-way systems costing much more, and that they have the added virtue *of simplicity*. It is probably fair to say that the better 2-way systems are almost as good as a 3-way system with similar components. One important difference is that in a 3-way system you have an extra handle on speaker balance by means of a control on the mid-range speaker. But, as advocates of 2-way systems would point out, this difference can be rather subtle and not easily noticed in casual listening. And in the best of the 2-way systems you can find some real winners.

In this chapter you will find four 2-way systems, three of which have wiring that is almost as simple as that of the single speaker systems in the preceding chapter. Two of them have ready-made crossover networks with coded leads; one has no crossover network at all. The other 2-way system uses a commercial crossover network that requires some alteration.

The size of the woofers varies from 5-¼″ in the smallest system to 10″ in the largest. Typical closed box 2-way speakers have woofers no larger than 8″ in diameter, but the 10″ woofer used in one of these has a specially treated cone for better mid-range response. While two of the speakers offer conventional styling, one is decidedly unconventional and another is designed to match Early American furniture. So look them over and see if one meets your speaker needs.

PROJECT 4: A CROSSOVERLESS 2-WAY SYSTEM

- **Construction**: Angled back, otherwise easy.
- **Wiring**: Simple. No crossover.
- **Comments** : Good range for size and cost, especially if given some bass boost from the amplifier(Fig. 4-1).

The development of *the piezo-electric tweeter* brought with it the possibility of a woofer/tweeter system that can divide the audio spectrum even though no crossover network is used. The impedance of a piezo tweeter is high enough that the amplifier isn't loaded by it. At low frequencies the impedance is so high that the tweeter has a natural roll-off, starting at a few thousand Hz. For

Fig. 4-1. Project 4 completed

best results at high power levels these tweeters should be used with a crossover network to limit the low frequency energy applied to them, but the crossover network must be especially designed for a piezo-electric tweeter.

Piezo Tweeter

One virtue of piezo tweeters used with no crossover network is superior transient response because there are electrical components in the tweeter circuit to prevent the amplifier from damping the tweeter. One frequently used setup consists of an 8″ woofer and a piezo tweeter in a compact box. A woofer of this size will normally have enough mid-range and lower high range to permit the use of such a tweeter. A problem with this kind of combination is that woofers are often relatively inefficient, requiring some kind of balance control for the tweeter. A simple capacitor is often put in series with the tweeter, not as a high-pass filter, but as a tweeter balance control. Typically the value of the capacitor may vary from 0.1 μf to 0.5 μf. The smaller the value of the capacitor, the greater the volume cut on the tweeter.

While a crossover may be desirable for these tweeters in large systems and a balance capacitor in smaller ones, it seems a bit wasteful to add a capacitor to a low-cost system with a piezo tweeter. It can be wasteful regarding amplifier damping as well as the capacitor's cost. A more economical way to achieve balance and to get wider dispersion at the same time is to install the tweeter on the rear of the cabinet. From that point no direct sound from the tweeter reaches the ear. Some of it is absorbed by wall surfaces, cutting the level, and reflections around the room give a non-directional quality to the treble that is more realistic than a direct-beaming tweeter. One disadvantage is that the tweeter must be placed near a reflective surface. A highly absorptive surface, such as heavy drapes, can muffle the highs. Except for this requirement a rear-facing piezo tweeter is an excellent choice, particularly for low-cost speakers.

The speaker system in this project was developed after considerable experimentation. Several careful listeners participated in the experiments. On one point their decision was unanimous; everyone agreed that the piezo tweeter gave a more natural sound when installed on the rear of the enclosure than when mounted on the front and controlled by a capacitor. Although this system was made rather small by building it around a combination of a piezo tweeter and a 5-¼″ woofer, it could be enlarged to a volume of about

0.75 cubic foot to accommodate an 8″ woofer and a piezo tweeter. If the box is enlarged for a larger woofer, it should be made from ¾″ material. Also it might be desirable to use one of the horn piezo tweeters instead of the lower priced version chosen here. Regardless of those speculations, the systems shown here gave good sound for its modest cost. That cost can be reduced even more if you have extra scraps of material handy or if you have access to the discard pile of a cabinet shop, a furniture factory or a building contractor. If the material you find is thicker than ½″, fine. Just enlarge the dimensions to maintain the same internal measurements.

Construction Procedure

Cut out the parts of the inner box according to Fig. 4-2. See Table 4-1 for materials. The back edge of the top and bottom pieces and the upper and lower edge of the back should be cut at an angle to match the slope of the back. That slope is about 72 degrees so, if you use a table saw, set your saw blade at 18 degrees. If you don't have the use of a power saw, you can estimate the angle with a hand saw and finish off the edges with a wood rasp or sandpaper.

Glue and nail the bottom to the sides. Then glue the top and the speaker board. Caulk the inner joints. Install the speaker leads on the back before putting the back on the box. You can use either an ordinary terminal strip inside the back for connections and route the cable to the receiver out through a caulk-filled hole; or you can use a

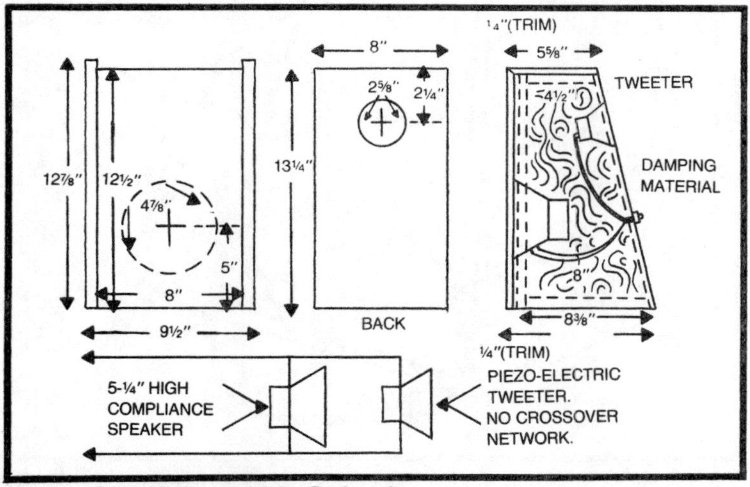

Fig. 4-2. Construction plans for Project 4.

Table 4-1. Materials list for Project 4.

	½″ Particle board, Pieces needed.	
1	8″ × 12″	Speaker board.
1	8″ × 13-¼″	Back
2	8″ × 12-½″ × 4-½″*	Sides. *Width at upper end.
1	4-½″ × 8″	Top
1	8-⅜″ × 8″	Bottom
	¼″plywood	
1	8″ × 12-½″	Faceplate
	¾″plywood	
2	9-⅝″ × 12-⅝″ × 5-⅝″*	Ext. sides. *Upper end.
	¼″× ½″ molding.	
1	48″ length, needed for trim, if desired.	
	Speakers:	
1	5-¼″ high compliance, Oaktron Model 525 CT or similar.	
1	2.5″ piezo-electric tweeter, Motorola KSN6016A	
	Speakers available from McGee Radio Co.	

two-screw terminal strip on the back. For the screw terminals you must drill two ¼″ holes to match the position of the terminals. Cut two 15″ lengths of lamp cord, one for the woofer and one for the tweeter (Fig. 4-3). Divide the leads at one end of each cord. Inspect

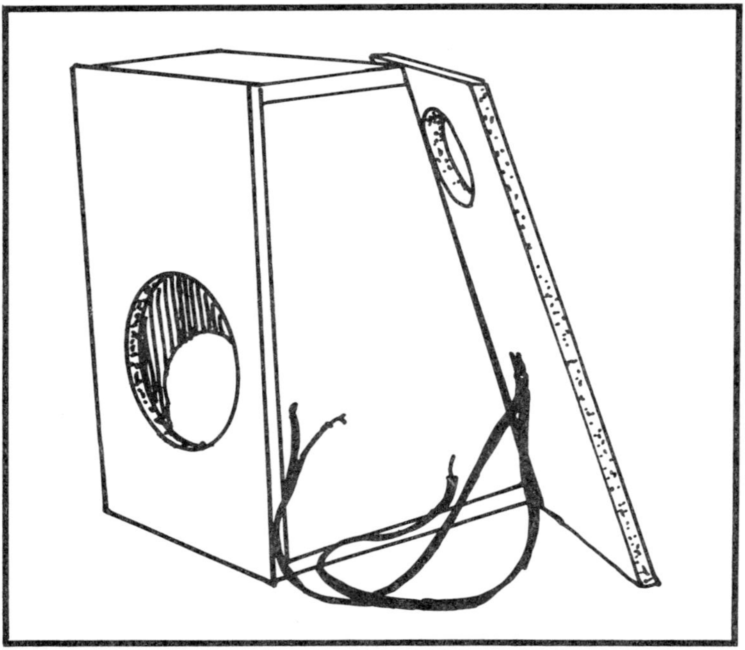

Fig. 4-3. Install two pieces of lamp cord before gluing and nailing the back to the enclosure.

each lead for beading, and pass the two beaded leads through one ¼″ hole, the two smooth leads through the other, from inside to the outside of the back. Connect the leads from each hole to a terminal on the strip and solder them. Install the strip on the outside of the back with wood glue, using tacks to hold the terminal strip on the back while the glue sets. Glue and nail the back on the box. When the glue sets, you can caulk the joints around the back.

An extra front panel board, with a cutout for the speaker, will hold the grille cloth flat. Cut a piece of ¼″ plywood to fit the front of the box. Set the woofer face down on it, in the identical position of the speaker hole in the box. Draw an outline of the speaker on the board and make the cutout. Lay the box on its back with a prop under the upper back to make the front level. Set the speaker in the box and check the fit of the front exterior panel or faceplate. Make any necessary adjustments in the speaker cutout or the outside edge dimensions; then paint it flat black.

Fig. 4-4. Plywood faceplate, glued and stapled or nailed to the speaker board, permits flush-mounted speaker.

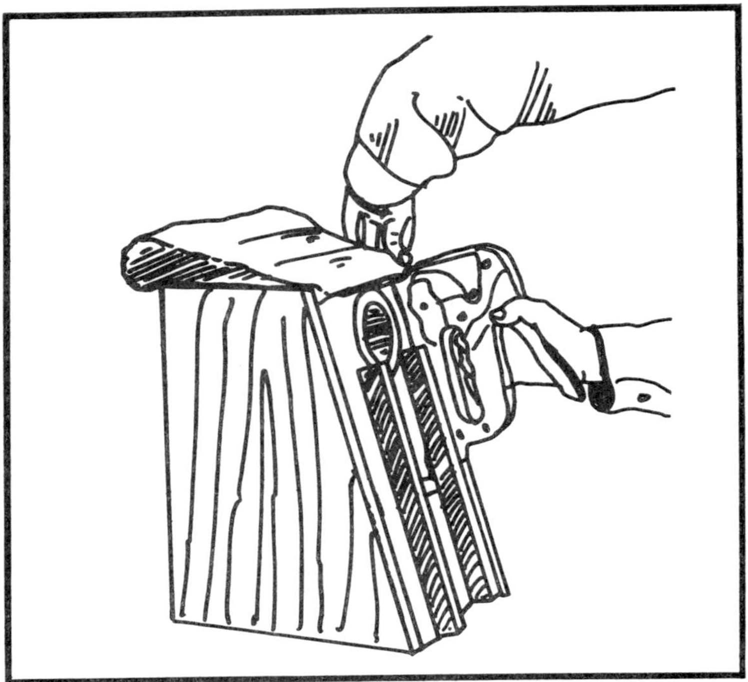

Fig. 4-5. Staple one end of grille cloth to the back above the tweeter, leaving the tweeter bare.

When the paint has dried, glue and staple or nail the faceplate panel to the front of the enclosure (Fig. 4-4). Cut out the external slab sides to cover the edges of the back, top and front with about ⅛″ to spare at the top and front. Although ¾″ material was used for sides in the enclosure pictured here, thinner material will work just as well and permit ¾″ trim to cover the edge of the grille cloth. Make a groove along the inner edges at the top and front of the side pieces to make space for the grille cloth.

Measure the exact length of trim pieces needed and cut them from solid wood. Sand and stain the sides and the trim. By putting the final finish on these pieces before installing the grille cloth, there will be no stains on the cloth.

Fill the enclosure with polyester fiber damping material. Feed the speaker leads through the damping material and out the holes. With no crossover components there is no need to distinguish between the separate lamp cord leads, but connect the beaded lead to the plus terminal in each case. The positive terminal on the tweeter is marked with a blue dot. Set the cabinet on its face, propped to make the back level. Wire the tweeter, solder the connections, and glue it down with a bead of silicone rubber around the edge of the hole. When the compound has set, invert the box and install the woofer. When propping the sloping back wall of the box, use care to protect the tweeter.

After about a half day you can test the speaker. If the test is satisfactory, install the grille cloth. Begin at the upper rear corner and staple one end of the cloth there (Fig. 4-5). Bring the cloth down across the front and staple it under the box. Stretch the cloth horizontally and staple it along each side of the enclosure. This installation leaves the tweeter uncovered for maximum high frequency response.

Install the trim on the edge of the side pieces by small brads. Or you can use 20 gauge nails that are ⅝" long. (Fig. 4-6). Drive them in until they hold, and then cut off the heads and drive them flush with the trim. Such nails are so small they hardly show.

Fig. 4-6. Cover raw edges of exterior plywood sides with pieces of ¼" × ¾" molding.

Fig. 4-7. Project 5 enclosure.

Glue a sheet of felt or plastic foam on the undersurface of the cabinet to protect any table top or other furniture upon which the speaker will be placed. Set the speakers near a wall or other reflective surface and connect them to your receiver. Move them to various angles and distances from the wall until you find the best location. When that is determined, this low-cost project is finished.

PROJECT 5: AN EXPANDABLE SYSTEM

- **Construction:** Use power saw.
- **Wiring:** Fairly simple. Commercial crossover adapted to use.
- **Comments:** Smooth quiet sound with extended highs in final system.

Here is a project that can grow on you (Fig. 4-7). You start with a single full-range speaker in a box. Then you add a tweeter, an L-pad and a crossover network. So if you don't have the cash handy

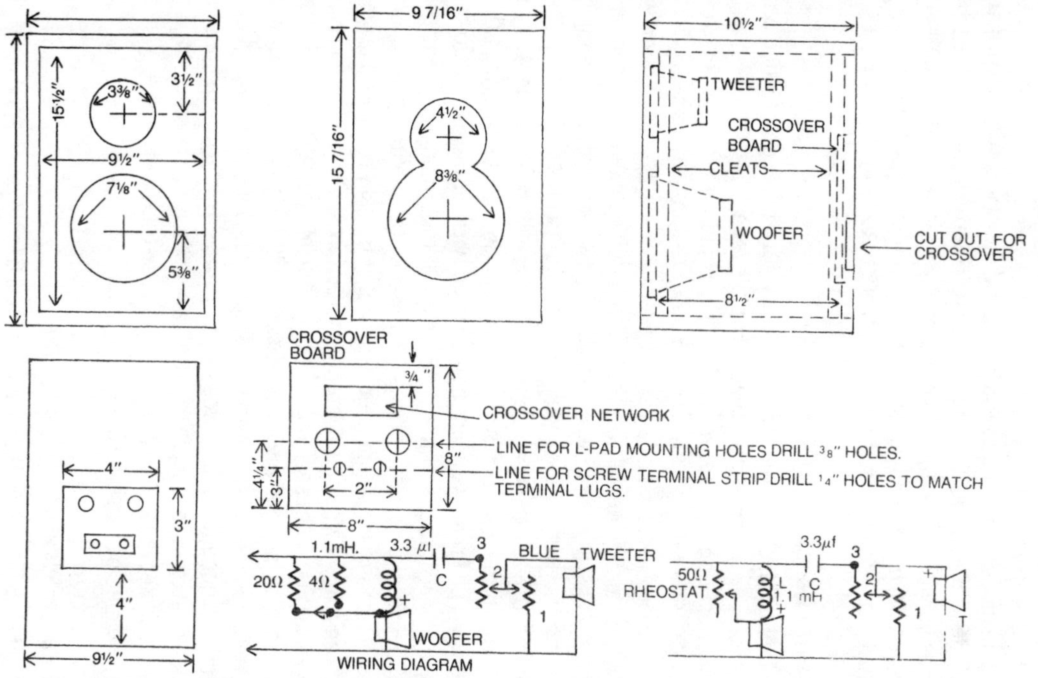

Fig. 4-8. Construction plans and wiring diagrams for Project 5.

for a full system, you can get started with just a couple of speakers. The project also has a couple of novel features that make it more interesting and improve the sound. See Fig. 4-8 and Table 4-2 for construction plans and materials.

The crossover hookup is unusual and deserves some explanation. These Radio Shack components were not designed for each other, but with a few modifications they work very well together and even have some advantages over a conventional hookup. One unusual advantage; a mid-range control on a 2-way system.

Crossover Network

Radio Shack's 2-way crossover network is recommended for any woofer-tweeter system that is suitable for a 2000 Hz crossover. Their small dome tweeter, used here, requires a crossover frequency of at least 4000 Hz. This tweeter is supplied with a 3.3 μf capacitor on one terminal so that the tweeter can be wired directly into any existing speaker system. If you build some boxes and put the 8″ speakers in them, you can add the tweeters alone by simply following the instructions you get with the tweeter. Then, if you later add the full crossover network, you will replace the 4.7 μf capacitor in the network with 3.3 μf capacitor that came with the tweeter. The crossover can then be used without further modification, but it will cause some depression of mid-range frequencies. This depression is not serious because of the gentle roll-off and because the full-range speaker, like most others, produces a bright sound anyway. If you add a potentiometer across the choke, you can vary its effectiveness from virtually full choke, when the total resistance is across the choke, to no choking action when the resistance is zero. The variable shunt resistance also swings the effective frequency of the crossover up and down. Thus the potentiometer acts as a control on the woofer's upper range. In the crossover network a switch and two resistors were used, which was cheaper but less flexible.

Full-Range Speaker

The project is built around a typical 8″ full-range speaker with an accordian suspension and a whizzer cone for extended highs. This kind of speaker offers the least expensive and simplest way to build a speaker system that approaches full range response. Almost all such speakers that are larger than 4 or 5″ in diameter have whizzers in the center of the cone to extend their range and improve dispersion of high frequencies. H. A. Hartley, the late founder of the

Table 4-2. Materials list for Project 5.

Speakers and Crossover Components		
1	8″ full range. Radio Shack #40-1341.	
1	Dome tweeter. Radio Shack #40-1276.	
1	2-way crossover network. Radio Shack #40-1293.	
1	Switch, Radio Shack #40-540.	
1	20 ohm resistor, 10 watts.	
1	4 ohm resistor, 10 watts.	
1	L-pad, 8 ohm. Radio Shack #40-980.	
Cabinet materials		
¾″ Birch Plywood		
2	10-½″ × 17″	Sides
2	10-½″ × 11″	Top & bottom.
¾″ Particle board.		
2	9-½″ × 15-½″	Speaker board and back.
⅜″ plywood		
1	9-7/16 × 15-7/16″	Grille frame
¾″ × ¾″ pine		
4	15-½″ long	Side cleats
4	8″ long	Top & bottom cleats.
¼″ Tempered masonite		
1	8″ × 8″	Crossover board.
¾″ wood veneer stripping		
1	60″ length	To cover plywood edges.

Hartley Company and a man who held strong opinions on how speakers should be designed, didn't like whizzer cones because of their free edge which, he said, added a "fizziness" to the sound. Hartley offered his opinion many years ago. Manufacturers of full-range speakers have ignored it because, as the representative of one large U.S. company reports, the whizzer does indeed extend the frequency range.

Careful listening and frequency response tests showed that the speaker used in this project, like all moderately priced speakers, had some peaks. The most notable one occurred at about 7000 Hz. Remembering Hartley's dislike of free cone edges, I decided to try some kind of damping on the whizzer. A few short strips of open cell foam weather stripping stuck to the inside edge made the sound subjectively smoother. A further check with a broadcast quality microphone indicated that the foam had indeed improved the response curve. So this is one modification that you can make for pennies.

Making Cutouts

If you decide to build the project by installments, you should make all the necessary cutouts for the tweeter and crossover

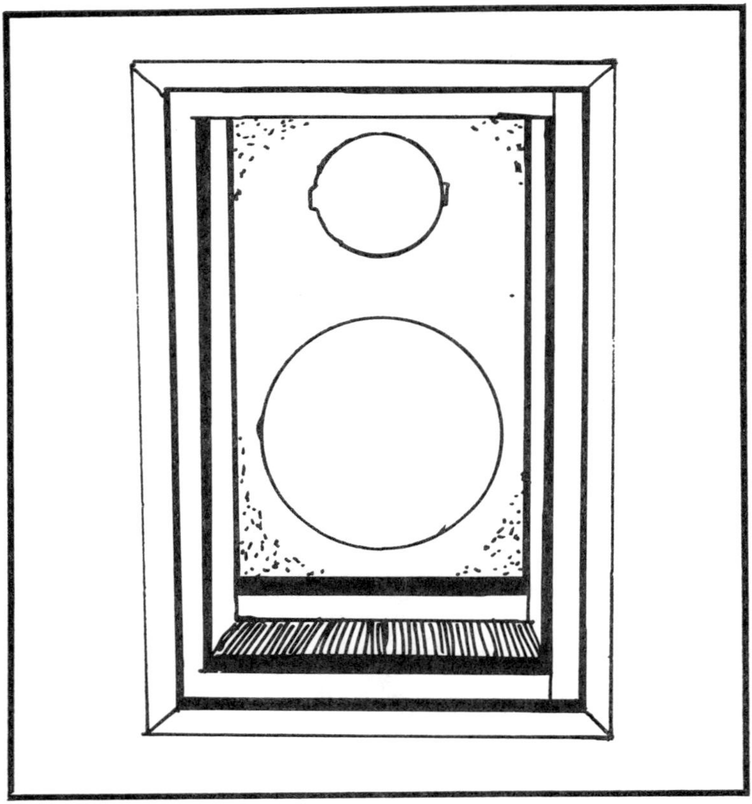

Fig. 4-9. Internal construction of box for Project 5.

network before assembling the box. You can use a piece of hardboard, plywood, or even a section from the woofer cutout to cap the tweeter hole. Just stick it in place with a bead of silicone rubber caulk. You can pry the cap off with a screwdriver when you are ready to install the tweeter. The crossover board can be installed on the back, minus the L-pad and crossover network which you can add later.

Birch plywood was used for the box shown in Figs. 4-9 and 4-10. You can substitute another material and butt joints if you prefer, but there was a reason for the hardwood plywood and bevel joints for this project. The method of construction shown permits all four sides to be finished. With butt joints, you will want to have access to the enclosure's interior for later installation of tweeter and crossover network. That is why this enclosure, unlike some other small boxes in this book, specifies cleats and a screw-in back.

Installing Cleats

A power saw is almost a necessity to make the 45 degree corner joints. If you don't have access to one, you can arrange for a cabinet shop to cut the parts to your specifications. When the sides are cut, you must decide whether to install the cleats on them before or after assembly. There are advantages to each method. The big advantage of installing them first is that you need only glue and nails for that job. The disadvantage is that it is difficult to make them fit perfectly because of minor variations in dimensions after the sides are glued together. One practical solution is to cut the cleats slightly shorter than the precise length of the space they will fill. If you leave no more than ⅛" gap, it can easily be filled later with caulking compound. To install the cleats after assembly, you must either use screws or clamps to hold them in place while the glue sets. Heavy nailing on the inside of the box is difficult in such limited space and can damage the joints.

Another decision that you must make before you start to work is the exact position of the front cleats. In the plan these cleats are recessed 1-¼", which is the correct total distance for an enclosure if the front edges of the plywood are covered by thin wood veneer tape.

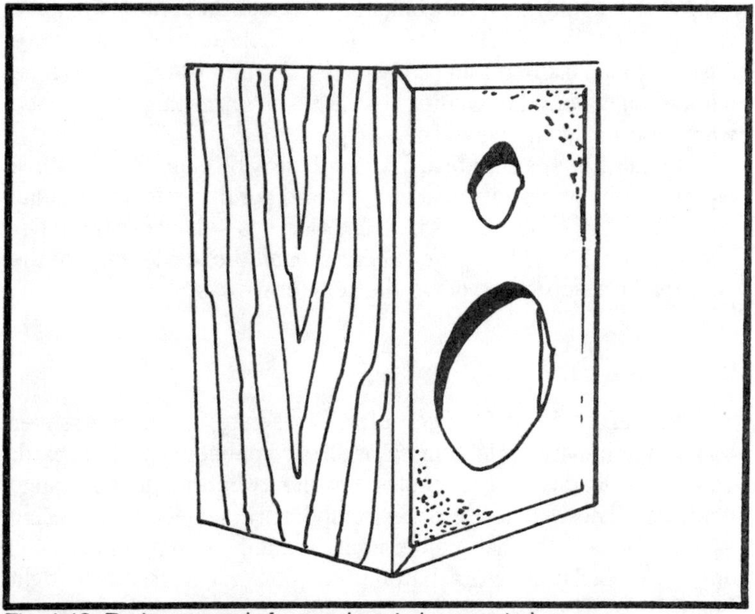

Fig. 4-10. Enclosure ready for speakers to be mounted.

Box Assembly

To assemble the box, put a coating of glue on each matching surface and tie the parts together with two tight straps or ropes. Or you can use two old 27″ bicycle inner tubes, each one circling the box twice. For the bicycle tube method you will need help in stretching the tubes and in keeping the box walls together while putting the tubes around them.

The chief problem at this state of construction is to avoid getting glue smeared onto the exterior walls of the box. Any glue there will seal the box against future staining. Use glue sparingly near the outer edges of the joints or cover the birch plywood near the joints with masking tape.

Cutting and Installing the Speaker Board

When you cut out the speaker board, you should make two notches opposite each other in the hole for the tweeter. If you have the tweeter on hand, you can simply cut the 3-⅜″ hole and rasp it out on each side until the tweeter will lie flat on the board. If you are preparing the box for later tweeter installation, you can approximate the correct hole shape by drilling two extra holes. First, mark a 3-⅜″ circle with a pencil compass. Then strike a horizontal line through the center of the circle. Drill a ½″ hole at each point where this horizontal diameter line crosses the circumference of the circle before you cut out the tweeter hole. Any minor rasping can be done when you get your tweeters.

Install the speaker board in the box with glue. You can use either clamps or screws to hold the board tightly against the cleats while the glue is setting. When the glue has set, caulk the inner joints. Run a bead of silicone rubber around the inside edge of the tweeter hole and press down the temporary cap.

Crossover Board

Prepare a 7″ × 8″ piece of ¼″ masonite for the crossover board. To simplify the later installation of components on this board, you can drill the two ⅜″ holes for the tweeter and mid-range controls. These holes can be temporarily sealed by installing ⅜″×½″ bolts with a flat washer under the nut on each bolt. A small amount of caulking material around the bolt will insure an air-tight board.

Drill two ¼″ holes to match the location of the solder terminals on the screw terminal strip you will use. Cut an 18″ length of lamp cord for the interior speaker cord. Remove about a half inch of insulation from each conductor and split the cord at one end to permit the leads to go through the holes in the crossover board. Pass the leads through the board and solder them to the lugs on the screw terminal strip. Install the terminal strip on the outside of the back with glue and small screws or tacks. Mount the crossover board on the inside of the back with silicone rubber sealer, making sure that the control positions are centered over the 3″ × 4″ cutout in the back.

Mounting and Hooking Up the Speaker

Mount the speaker from the front of the box with silicone rubber. If you have the tweeter on hand, you can mount it too, but it's a good idea to first solder the leads of an 18″ length of lamp cord to its terminals. This operation can be performed after the tweeter is installed, but the limited space around the lugs makes it difficult.

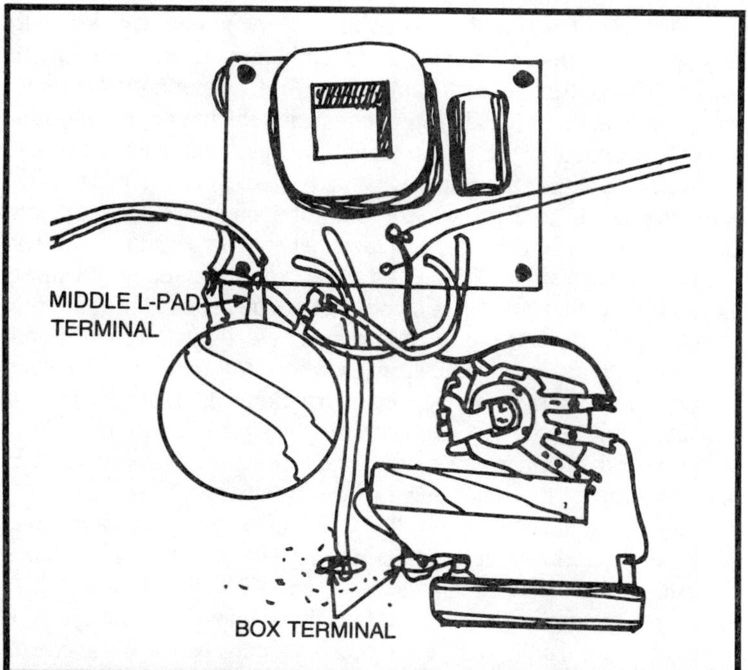

Fig. 4-11. Crossover wiring. Woofer lead goes right and the tweeter lead is from the L-pad to left.

Make a note of which lead in the cord goes to the positive terminal of the teeter.

You can now hook up the speaker by running the lamp cord from the terminals on the crossover board to the speaker lugs and soldering them. To insure proper speaker phasing, connect the lead from the left speaker box terminals, as seen from outside the box, to the positive terminal on the speaker. This wiring corresponds to the polarity of the hookup shown in Fig. 4-11 of the completed crossover network.

Finishing the Enclosure

Fill the enclosure with loose polyester damping material. Make a gasket for the back by carefully cutting four pieces of 3/8″ × 1/4″ open cell foam weather stripping material and installing it around the cleats just inside the line of screws. Or use a narrow bead of mortite. Use care to see that your gasket material doesn't get picked up by the screws and twisted out of shape or broken; that is why it must be placed either inside or outside the screw line.

The foam treatment of the whizzer cone is optional. While it may give the effect of slightly reduced high frequency output, it makes the sound subjectively smoother. And, as mentioned earlier, tests made on the original system confirmed the increased smoothness of response. If you plan to apply the foam, you should choose lightweight weather stripping; what is needed is enough material to damp the free edge of the whizzer but not enough to overload it and kill the upper highs (Fig. 4-12). Open cell polyfoam that is no wider and thicker than 3/8″ × 1/4″ is about right. You will need only a short piece for this job, 8″ or 10″, so you can use the remainder of the roll to seal the back. Cut the foam into short pieces with tapered end cuts so there are no significant gaps where the ends meet. If you imagine the foam is the outer ring of crust on a pie that is to be cut radially, you'll get it right. Remove the backing and stick the adhesive side of the foam inside the rim of the whizzer, using your other hand to support the outer edge of the whizzer.

Make a grille board according to the plan and stretch any open weave grille cloth over it and staple it on the back edges. If you make the grille board to fit carefully, a fraction fit will hold it in place. If it is loose, you can apply pieces of velcro binding material to the speaker board at several points and put matching pieces on the back side of the grille board. This completes the enclosure construction and the first installment of your speakers.

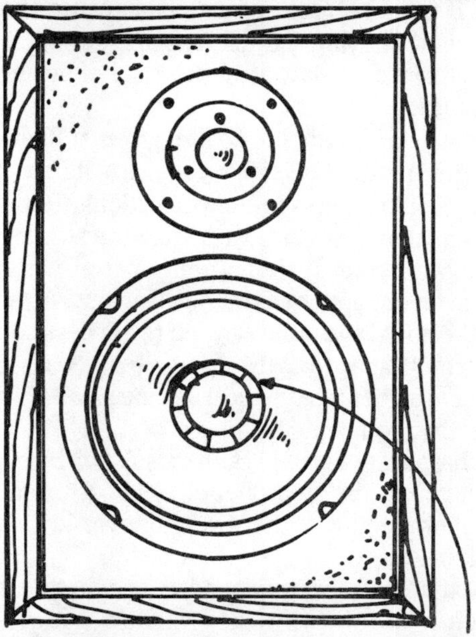

Fig. 4-12. Small pieces of lightweight foam weather stripping are attached to the inner lip of the whizzer cone. The foam damps the free edge of the whizzer and smoothes its response.

Tweeter and Crossover Wiring

You can add a tweeter by simply wiring it in parallel with the full-range speaker, but you must wire one tweeter lead to the outside lead of the capacitor that is mounted on the tweeter. While this simple maneuver will extend the high frequency response of the system, it will also unbalance the sound in favor of the highs, at least in most listening environments. With such a combination you will be tempted to make excessive use of the treble cut control on your receiver, killing the upper high response that you wanted to get when you bought the tweeter. An L-pad is almost a necessity when you add the tweeter. Unlike the treble control on your receiver, an L-pad is non-frequency discriminating. It will reduce the tweeter's output at every frequency reproduced by the tweeter instead of rolling off its upper range.

ADDING A CROSSOVER NETWORK

The final step in this project is to add a full crossover network, which limits the high frequency energy to the woofer. This has two advantages. It reduces intermodulation distortion that is caused by a single speaker carrying the full frequency range, and it gives a unique opportunity for controlling the level of sound output in the upper mid-range.

The crossover network is adapted from a Radio Shack 2-way crossover, the 40-1293. Before you wire it into your speaker system, you should make some modifications. Remove the 4.7 μf capacitor from the crossover board. Heat the solder on the foil side of the board with your soldering gun or iron while grasping the capacitor lead on the upper side and pulling out. After the capacitor is out, heat the board again and sling off the excess solder so you can install the correct capacitor. Also, remove the 3.3 μf capacitor from the tweeter and solder it to the board in the same position as the 4.7 μf capacitor that you removed.

Use a sharp knife to cut back the insulation from the ribbed red lead where it enters the crossover circuit board. The purpose here is to bare a short section of the wire, not over ¼″, so you can later solder another lead to it.

This next step will vary depending on whether you want to go to the expense of a continuously variable mid-range control or stepped control. For a variable control you will need a 50-ohm potentiometer or *rheostat*. For example, an Ohmite 0149 rheostat would be just right, but it costs at least twice as much as the 3-way speaker switch and resistors.

If you choose the cheaper switch, pull off the control knob and mount the switch in its proper hole on the large crossover board inside the back. You will probably have to countersink this switch to get the retaining nut to catch the threads. Mount the crossover network at the same time by squirting a bead of silicone rubber sealant under each of the four mounting holes and pressing it down on the larger board. Wait until the sealant sets before proceeding.

Cut the yellow lead on the speaker switch to 3″ long. Remove the insulation from the end of it and solder it to the section of the ribbed red lead on the crossover board that you have already laid bare.

Remove the red lead from the speaker switch. Solder one lead of a 20-ohm resistor to the lug that held the red lead.

Remove the black lead from the switch. Solder one lead of a 4-ohm resistor to the lug that held the black lead.

Remove the speaker leads that you have been using for the full-range speaker from the box terminals. Cut off the black leads from the crossover network just long enough to reach the box terminals. Bare the leads and connect them to the terminals, preferably with the ribbed black lead going to the lug nearest the speaker switch.

Connect the free ends of the two resistors to the same lug that has the ribbed black lead connected to it. Solder the connections at both lugs. Remove the tweeter leads from the L-pad. Cut the blue leads from the crossover network to the proper length. Connect the ribbed blue lead to lug #3 of the L-pad, the smooth blue lead to lug #1. Connect the lead that goes to the positive terminal on the tweeter to lug #2 of the L-pad and the negative lead to lug #1. Solder the connections at the three terminals of the L-pad.

Connect the leads from the crossover network to the woofer. The ribbed lead should go to the positive terminal of the woofer and the smooth lead to the negative terminal.

This completes the wiring of your modified crossover network and speaker system. Install the back temporarily and check the sound at low volume to see if the highs come from the tweeter and the lows from the woofer. Check the sound with the mid-range switch in its various positions. As you twist this control to the last counterclockwise position, there should be a significant drop in mid-range level. There is less difference in the middle position and the most extreme clockwise position because the latter position puts the 20-ohm resistor in parallel with the 4-ohm resistor, giving a net resistance of 3.33 ohms.

When you are sure the speaker system is correctly wired, install the back permanently. You'll find this speaker system delivers very smooth sound or as one listener described it, "harmonious."

PROJECT 6: THE COLONIAL

- **Construction:** Easy. Can use hand tools.
- **Wiring:** Simple. Pre-wired crossover, slip on connectors.
- **Comments:** Good range with slightly plump bass.

If you want a speaker that will match your Early American furniture, this is it (Fig. 4-13). The enclosure for this system was designed to fit the dimensions of a shutter panel which can be used as a grille frame. Shutter panels come in various sizes; the only hitch was to find a shutter panel that would fit the front of a box with just the right volume for a suitable speaker system (Fig. 4-14). The final choice was a 12″ × 20″ shutter panel on a pine-covered box with 0.7 cubic foot volume that encloses a Speakerlab 1 system. The speakers are available as a kit of woofer, tweeter and crossover network.

Fig. 4-13. Pine wood box and shutter panel grille frame give an early American look to Project 6.

You can build the box and install a kit without even a soldering iron. Speakerlab puts push-on connectors on the coded leads from the crossover network, making wiring quick, easy and virtually foolproof.

This system gives exceptionally wide range sound for the size of the enclosure, although a few listeners considered the bass a trifle heavy. Speakerlab has made some changes in their speaker line recently, but their new S-1 system will work equally well in this enclosure because the box requirements are identical to those of the components shown here. If you want to build a good 2-way system but dislike Early American furniture, you can make a plain grille frame and save the cost of the shutter panel. Some shutter panels add a slight coloration to the otherwise open treble of the Speakerlab tweeter, so you may want to install hinges on the panel and swing it open while using the speaker. But remember this, most listeners won't notice the slight difference in sound introduced by the panel.

Box Construction

The enclosure for this speaker is made from solid pine plus a particle board liner. See Fig. 4-15A and Table 4-3 for plans and materials. This combination makes for easy construction with more than adequate wall strength and rigidity. You can cut the 9-¼″ wide pine parts from 1″ × 10″ pine boards because lumber sold as 1″ × 10″ is actually only 9-¼″ wide. The dimensions of the box are designed to match this available width so you don't have to rip any boards. A hand saw is all you need to cut out the individual parts. Measure the shutter panel to get the exact dimensions you need for the front of the box. Some shutter panels that are sold as 12″ × 20″ panels will vary from these measurements by ⅛″ or so. Don't worry about this affecting the performance of your system; the minor change in dimensions will have a negligible effect on internal cubic volume.

Cut out the pine pieces for the sides, top and bottom. Glue and nail these parts together, using #6-penny finishing nails, to make the outer shell of the enclosure. Next, cut out the pieces of ½″ particle board that fit inside the top and bottom of the shell. Glue and clamp these parts inside the enclosure (Fig. 4-16). Or, if you don't have clamps, nail them in place with nails that are 1″ long. If you use

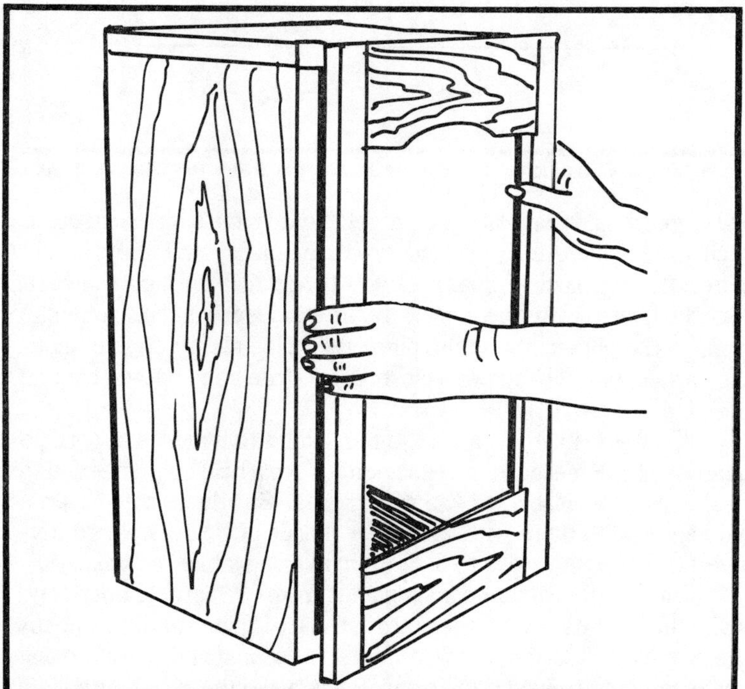

Fig. 4-14. Pine box should fit the dimensions of the shutter panel.

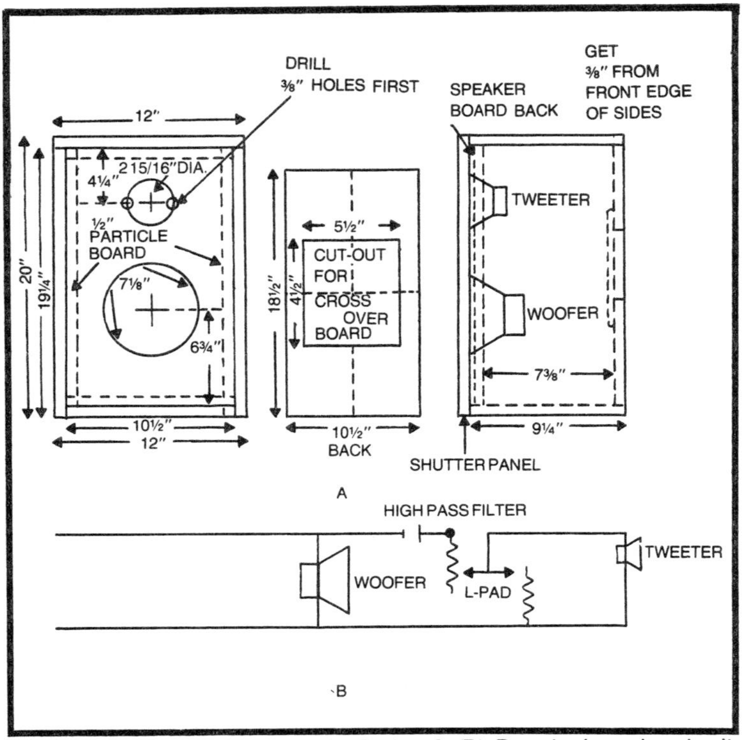

Fig. 4-15. (A). Construction plans for Project 6. (B). Pre-wired speaker circuit.

nails, you should first place the particle board panels on a solid base, such as a concrete floor, and drive the nails until they barely penetrate the particle board. If you try to nail through the hard particle board with the board in the enclosure shell, you may damage the glue joints on the pine walls. Use plenty of glue under the particle board to insure a close, rattle-free bond between it and the pine.

When you have installed the top and bottom pieces of particle board, you can measure the exact length of particle board needed to fit the space available for the side pieces. Cut these pieces; then glue and install them with clamps or pieces of board wedged between the opposite walls to hold them in place while the glue sets.

Cut out the front and back panels from ¾" particle board and make the cutouts for the speakers in the front panel and the one for the crossover network in the back. Glue and install the back, using heavy weights to hold it under pressure while the glue sets. Then you can sand, stain and varnish the pine exterior of the box and the

Table 4-3. Materials list for Project 6.

1″ × 10″ pine boards, actually ¾″ × 9-¼″, pieces needed.			
	2	9-¼″ × 19-¼″	Sides
	1	9-¼″ × 12″	Top
	1	9-¼″ × 10-½″	Bottom
½″ Particle board for liner, pieces needed.			
	2	7⅜″ × 10-½″	Top & bottom, cut to fit.
	2	7-⅜″ × 17-½″	Sides, cut to fit.
¾″ Particle board, pieces needed.			
	2	10-½″ × 18-½″	Speaker board and back.
Also need one 12″ × 20″ pine shutter panel.			
Speakers and components:			
	1	Sls speaker kit, 8″ woofer, DT 108 tweeter and X1 crossover. Kit available from Speakerlab.	

shutter panel grille frame. If desired, you can paint the speaker board flat black. This is also a good time to seal the interior joints of the enclosure with latex or silicone rubber caulking compound.

Crossover Network

When the cabinet is finished, install the crossover network. Set the enclosure on its back and run a bead of silicone rubber around the cutout in the back. Press the crossover board down over the cutout, making sure the speaker terminals and tweeter control

Fig. 4-16. Glue particule board liner in box shell. Clamp the liner or wedge pieces of wood between opposite sides until the glue sets.

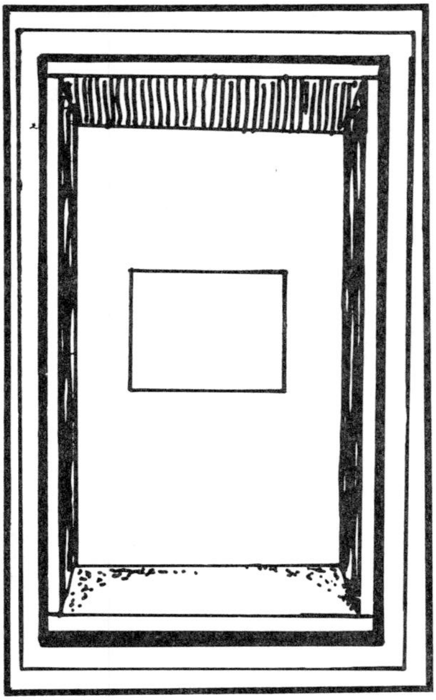

Fig. 4-17. The box is ready for installation of crossover board over the window in back.

are centered in the opening (Fig. 4-17). Also check to see that the tweeter control will be above the speaker terminals when the box is set upright. Leave the box in that position for at least several hours while the sealer sets.

Cut a piece of fiber glass or polyester damping material to cover the interior of the back. Make a slit reaching from one edge of the fiber glass to the center. Pass the speaker leads through this slit, and then staple the fiber glass to the back. The most desirable thickness of the damping material will depend on its density. A thickness of about 2″ to 3″ is probably adequate. Cut pieces of damping material to fit on the interior of the top, sides and bottom and install them.

Next, install the speaker board. Run a bead of silicone rubber sealant around the front edge of the particle board liner; then press the speaker board down firmly on the sealant. Use care in doing this to insure that there are no gaps in the rubber sealant. The bead of silicone rubber should be placed near the outer pine wall so it will fill any recesses there. Stain and finish the box (Fig. 4-18).

Wiring the Speakers

Next wire the speakers (Fig. 4-15B). The leads to the woofer and tweeter are color coded, with the blue lead going to the positive woofer terminal and the yellow lead going to the positive tweeter terminal. The negative leads are cloded blue/black for the woofer, yellow/black for the tweeter. If you should get a kit with two black leads, these will be identical electrically but will have push-on terminals of different size to match the terminals on the woofer and tweeter. Separate the tweeter leads and the woofer leads and pass them out through the appropriate speaker holes. Connect each lead to its matching speaker terminal by gently pushing it on. You may have to rock the push-on terminals to make them go on. Rock in the direction of the wider dimension of the speaker terminal, not in the direction that will bend and weaken the terminal. Make sure that the tabs are pushed all the way onto the terminals.

Set the speakers face up in their holes and prop the box across an empty carton or other prop so you can reach the control at the back. Make sure the power is off on your receiver. Turn the volume control all the way down. Connect the speaker to the receiver and turn on the receiver. Turn up the volume control just enough to easily hear the speakers. Listen to them to see if the bass comes from the woofer and the treble from the tweeter.

Adjust the tweeter control to see if it is working properly. Don't expect balanced sound at this stage because the speakers are not properly installed in the box and you will lose the deep bass. And

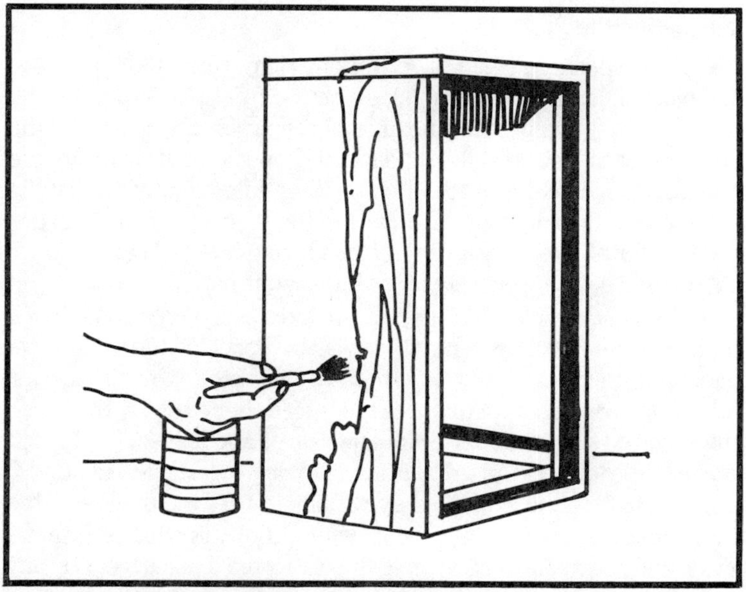

Fig. 4-18. Stain and finish box before installing speakers.

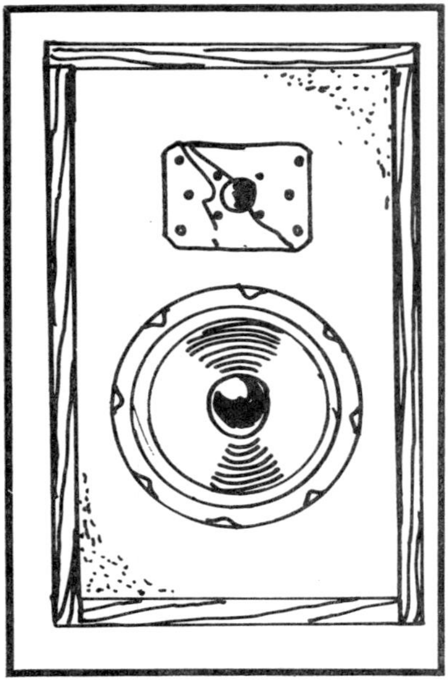

Fig. 4-19. Location of speakers on board. Latest kit from Speakerlab may include a new model of tweeter, but box requirements are the same.

above all don't be tempted to turn the volume up high while the speakers are loose.

Completing the Project

If the speakers seem to be working right, turn off the receiver and disconnect it. Set the speakers on the front panel to leave the holes free for gluing. Run a bead of silicone rubber glue around the edge of each speaker hole. Place the speakers down into the mounting holes (Fig. 4-19). Rotate them just slightly to make a proper seal; then press them down tightly. Leave the cabinet in this position for at least a half day before moving it or using it. Allow a day or more drying time before you drive it hard.

Cut a piece of grille cloth to fit the depressed area on the back of the shutter panel. Measure the available space on the back of the panel and then add about ½″ to each dimension. On the shutter panel shown in Fig. 4-14 this area measured about 9-¾″ × 15-⅝″, so the required size of the grille cloth measured about 10-¼″ × 16-⅛″. Staple the cloth on the grille frame, driving the staples in at a 90 degree angle to the cloth so they will aim toward the outside of the frame instead of toward the front, where they might penetrate the wood and show. After stretching the cloth until it is taut, cut off any excess tufts.

Test the speaker with the grille frame held before it and with the grille removed. The reason for this test is that shutter panels of slightly different design may show different degrees of sound coloration. If you hear no coloration, you can attach the shutter panel with a small amount of silicone rubber glue well inside each corner. This will permit you to remove the panel at any time by prying it off. If your shutter panel does cause an unacceptable degree of coloration, you can either make a plain grille frame, such as is described in other projects in this book, or put small hinges and a cabinet catch on the shutter panel. Regardless of the final treatment this easy-to-build compact system will deliver a lot of sound.

PROJECT 7: THE MAXI 2-WAY

■ **Construction:** Power saw. Or make particle board box with hand tools.

■ **Wiring:** Simple. Pre-wired crossover with push-on connectors.

Fig. 4-20. Special woofer cone treatment permits more extended range and makes 2-way system possible with 10″ woofer.

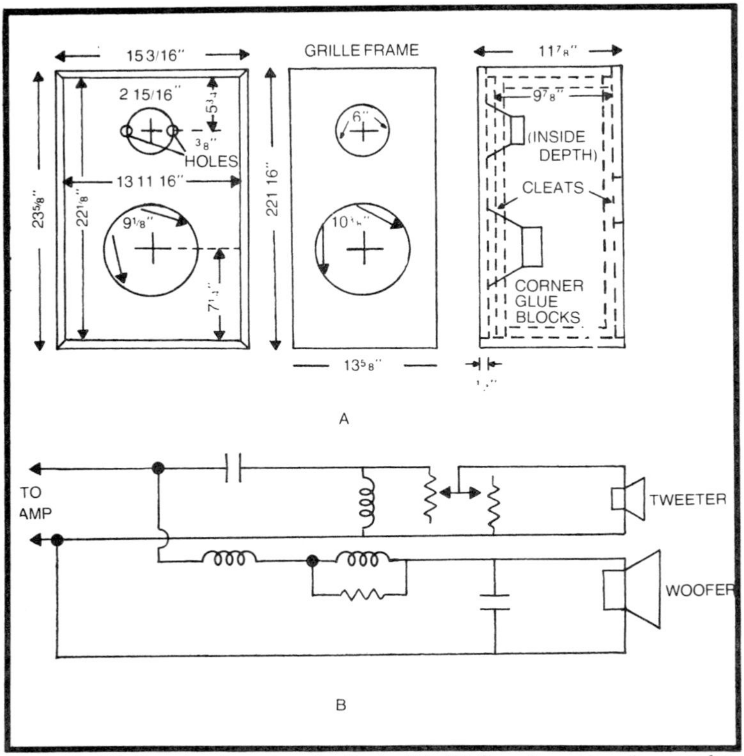

Fig 4-21. (A). Construction plans for Project 7. (B). Speaker wiring diagram for Project 7.

■ **Comments:** Full sounding 2-way with considerable presence. 4-ohm impedance. Bass well damped.

Most closed box speakers with 10″ woofers are 3-way systems. Considering the response characteristics of most available 10″ woofers and of small tweeters, a 3-way system is almost mandatory to avoid problems in mid-range reproduction. The problems with larger 2-way speakers is that the woofers cut off at too low a frequency to put the crossover point above, say, 1000 Hz, and very few small tweeters can reproduce frequencies that low without adding peaks or distortion to the music. A larger tweeter can be used, but in that case the upper end response and dispersion would suffer. Speakerlab solves the problem by special treatment of the cone on their 10″ woofer so the crossover point can be placed at 2000 Hz (Fig. 4-20). This permits the use of a small dome tweeter to reproduce the upper mid-range and high frequencies.

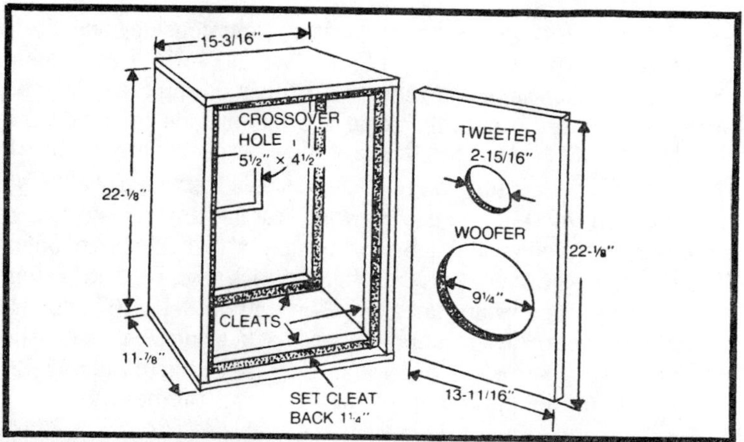

Fig. 4-22. Construction plans for an alternate enclosure for Project 7, using ¾" particle board.

The speakers used in this project are sold in a series of kits consisting of speakers and crossover only: speakers, crossover, and a utility enclosure; speakers, crossover, and walnut veneer enclosure; or speakers, crossover, and a finished solid walnut enclosure. If you buy only the speakers and crossover network, you can build a cabinet and put any kind of finish you desire on it. With that in mind two sets of plans are shown here, one with butt joints, the other with beveled corners (Figs. 4-21A and 4-22 and Tables 4-4 and 4-5).

Table 4-4. Materials list for Project 7.

¾" Birch plywood, pieces needed.		
2	11-⅞" × 23-⅝"	Sides
2	11-⅞" × 15-3/16"	Top & bottom
¾" Particle board, pieces needed.		
2	13-11/16" × 22-⅛" Speaker board and back.	
⅜" Plywood or particle board, one piece needed.		
1	13-⅝" × 22-1/16"	Grille frame.
¾" × ¾" pine, pieces needed.		
4	22-⅛" long	Side cleats
4	12-3/16" long	Top & bottom cleats.
¾" Wood veneer stripping.		
1	78" long	Veneer for front edges of plywood walls.
Speakers and Components:		
1	S2s kit, including W1004B woofer, DT108 tweeter. X2 crossover. Available from Speakerlab.	

Before you begin construction, you might consider one alternative to the two sets of plans in Figs. 4-21 and 4-22. If you want a pine cabinet, such as the one shown in Project 6, you can follow the plans for that project, but change the dimensions to obtain the internal volume needed for this larger system—slightly over 1.5 cubic feet. The enclosure in Fig. 4-20 was built from high quality ¾" hardwood plywood. If you use plywood that has any voids or other imperfections, you should consider using a ⅜" particle board liner. If you do that, you should add ¾"length to each side, top and bottom panel. The particle board liner pieces would be 9-15/16" wide. It would fit between the speaker board and back, so no cleats would be needed. To add particle board you can cut the sides to their 11-⅞" width, glue on the particle board and then cut the bevels.

Cutting Box Parts and Cleats

To build the enclosure, begin by cutting the sides, top and bottom to width. Set your saw to make a 45 degree end cut and check it by cutting off two pieces of waste material. Set the two pieces together and check the corner they make with a square. The corner should be square with no gaps in the joint. When you are satisfied that the saw is set correctly, cut the sides to proper length.

Cut four cleats to the 22-⅛" length. Glue and nail these cleats onto the interior surface of the side walls. The rear cleats should be recessed from the edges by ¾".

Coat the beveled ends of the sides with wood glue and assemble them on edge on your work bench. Use buckled straps or elastic straps to hold the parts together while the glue sets. During this part of the procedure you must use great care to prevent glue from getting on the plywood exterior. It's a good idea to get someone to help you hold the parts together while you glue the joints and put the straps around the enclosure shell.

Cut four short cleats to fit exactly in the space available for them between the side cleats already installed. Glue and clamp or screw these cleats to the interior of the top and bottom, using #8 × 1-¼" screws. Cut out the speaker board and the back, making any minor adjustments in dimensions so they will fit the space available. Make the cutouts for the speakers in the speaker board and for the crossover board in the back. After the glue holding the cleats in the enclosure has thoroughly set, glue and nail the back board into the box. After the glue on the back has set, caulk the interior joints with silicone rubber glue or latex caulking compound. Then place the box

Table 4-5. Materials list for the alternate enclosure for Project 7.

3/4" Particle board, pieces needed.	
2 15-3/16" × 11-7/8"	Top & bottom
2 22 1/8" × 11-7/8"	Sides
2 22-1/8" × 13-11/16"	Speaker board and back.
1/2" plywood or particle board. One piece needed.	
1 13-5/8" × 22-1/16"	Grille frame.
3/4" × 3/4" pine, pieces needed.	
4 22-1/8" long	Side cleats
4 12-3/16" long	Top & bottom cleats.
Plastic veneer.*	
1 13-1/2" × 78" long	Covering material.

*If desired, 1/4" birch plywood can be used as veneer. See use in other projects.

on its back and run a bead of silicone rubber glue around the edge of the cutout in the back. Press the crossover network down over the glue, checking to see that the speaker terminals and the tweeter control are centered in the opening. Leave the cabinet in that position for a half day or more while the glue sets.

Attaching Damping Material and the Speaker Board

Cut a piece of fiber glass or polyester batting to cover the interior of the back. Make a slit from the edge to the center of the crossover network location. Feed the speaker leads through this slit and staple the damping material to the back. The thickness of the damping material will depend on its density. It can vary from about 1" thick for dense fiber glass to 2" or 3" for lighter material. Cover the sides, top and bottom with a layer of fiber glass.

You can paint the speaker board flat black if desired. Install the speaker board, using plenty of glue. Inspect the cleats where the speaker board will be installed. Fill any voids with caulking material. If there are no voids, you can use wood glue and nails to install the front board. But if you have any doubts about the ability of wood glue to seal the box, you should install the speaker board with silicone rubber.

Sand and stain the cabinet. Follow this with an oil treatment or a varnish finish as detailed in Chapter 1.

Connecting Wires to Terminals

Feed the speaker leads out through the appropriate speaker holes (Fig. 4-21B). The yellow and yellow/black striped wire goes through the tweeter hole. The blue and blue/black striped wire goes through the woofer hole (Fig. 4-23). Connect the wires to the

Fig. 4-23. Crossover is pre-wired with color coded leads.

speaker terminals by pushing the connectors onto the terminals. Push the connectors on straight or by rocking them slightly. But don't rock them in the direction that will bend the speaker's terminals. The yellow wire goes to the tweeter's positive, or red dot, terminal; the yellow/black wire to the other tweeter terminal; the blue wire to the woofer's positive terminal; and the blue/black wire to the woofer's other terminal.

Set the speakers in the holes temporarily. Carefully prop the cabinet high enough so you can reach the terminals on the back. Check to see that your receiver is turned off. Turn the volume control all the way down. Connect the receiver to your speaker system and turn on the receiver. Turn up the volume just enough so you can easily hear the speakers. Listen to see if the lower frequencies are coming from the woofer and the highs from the tweeter. Adjust the control on the back to see if it is working properly. Don't drive the speakers hard before they are installed in the box. And don't expect well balanced sound; the low bass will be lost from the virtually unbaffled woofer. When you are satisfied that everything is working, turn off the receiver and disconnect the speaker.

Installing the Speakers

Next install the speakers. Set them aside on the speaker board and run a bead of silicone rubber around each speaker hole. Set the speakers into position and rotate them just slightly to insure a complete seal. Press them down hard and leave them for a half day or more before moving the cabinet.

Cut a grille board from 3/8″ particle board or plywood. Paint it flat black, if desired. Cut a piece of grille cloth to fit over the board. Stretch the cloth over the board and staple it on the rear edge. This completes the project.

When you order your speakers, you will get a set of enclosure plans from Speakerlab that shows the same volume as this one. But it is slightly different in construction and styling. The sound will be the same no matter which one you build.

3-Way Closed Box Speakers

In the early days of high fidelity there were very few 3-way speaker systems, but since the acoustic suspension revolution the 3-way has become the norm. The stampede to 3-way systems occurred because acoustic suspension woofers have heavier cones with longer voice coils than earlier woofers. These both contribute to a reduced upper frequency range. If the woofer is larger than 10″ in diameter, a crossover point of 1000 Hz or lower is usually necessary for adequate mid-range reproduction. Such a crossover frequency is easy to obtain in a 3-way setup, but it is much more difficult for just a woofer and tweeter.

THE FACTS ON 3-WAY SPEAKERS

Another reason for going to a 3-way system is that you have an extra control in the mid-range pad, a way to shape the total response of your system. This does not mean that a 3-way system is automatically better than a 2-way system with similar components. Many engineers who advocate 2-way speakers contend that the more complex the system, the greater the chance that components will interact unfavorably or that something will go wrong. As the controversy goes on some audio people insist on 2-way speakers; others demand 3-way systems. A more practical approach is to judge each speaker on its own merits without regard to preconceptions about which is best. Here are four 3-way systems that went together with very few problems and, in their final version, gave excellent performance.

In size these speakers vary from a compact bookshelf model to a full-sized floor speaker. All of the woofers were designed for closed box operation, but one, used in Project 10, was considerably more efficient than the others. That might be a consideration if you have a receiver with marginal power output ability.

In difficulty of construction, Projects 9 and 10 require a power saw if you follow the plans exactly, but you can adapt those projects to hand tools by using butt joints for the pine enclosure. Three of these projects use commercial crossover networks. Project 10 requires a homemade crossover. But you can get almost as good performance from an across-the-counter crossover. More about that later. If you have wanted to build a 3-way speaker and have been intimidated by crossover problems, now is the time to put your reservations aside and go to work.

PROJECT 8: A COMPACT 3-WAY

- **Construction:** Fairly easy.
- **Wiring:** Fairly simple. Commercial crossover.
- **Comments:** Controlled wide range for size.

Most speaker systems designed around an 8″ woofer consist of only two drivers, the woofer and a tweeter. Woofers of this size usually have good response to 2000 or even 3000 Hz, so there is a wide choice of tweeters to handle the upper range. So why a 3-way here? The reason is chiefly because listening tests showed this combination to be a winner. One probable reason is that the small mid-range cone offers good dispersion. And raising the crossover point to the tweeter permits the use of a super tweeter that can carry the highs to the limits of audibility and beyond.

Another reason for the 3-way hookup is the previously mentioned opportunity to shape the full response of the system by means of the mid-range control and tweeter control. You can get similar balance in a 2-way system, but you may have to tinker with the crossover network to do it. Here it is done the easy way.

For a 3-way, this system is easy to build and wire. The only unusual construction step is to build a mid-range enclosure to isolate the open back mid-range driver. As for wiring, you will have to add one component, an extra resistor across the super tweeter to set the crossover frequency. If you need a compact speaker for a small to average size room that covers a wide frequency range with ease, try this one.

Speaker Components

The woofer is the Radio Shack 40-1006, with a 16-ounce magnet and a foam suspension. The mid-range driver, Radio

Shack's full range 4″ speaker (40-1197), has been continued in stock for many years. A question may be raised here. Why wasn't a driver designed for mid-range use chosen? In this case the driver selected was considered better suited to the job than any mid-range driver of similar cost. Space on the box was too limited for a larger driver, and a larger driver would have poorer dispersion. Also, many low-cost mid-range drivers have a bad bump in their response curves, bumps that are too high in frequency for the crossover network to eliminate them. Here the mid-range box volume, generous for a mid-range driver, is large enough to permit a resonance that is far below the crossover frequency. That resonance is also well damped by the 7-ounce magnet on the little 4″ speaker.

For high frequency duty a Radio Shack 40-1280 piezo-electric horn super tweeter gives response into the ultrasonic range with excellent transient response.

The crossover network is Radio Shack Catalog Number 40-1339A (Fig. 5-1B). Note that this model has different crossover frequencies from the older 40-1339. With 8-ohm speakers, the "A" model crosses over at 1000 and 6000 Hz, which are ideal choices for the drivers used in this project. The piezo-electric super tweeter can be used without a crossover network, but with one small change it can be wired right into the crossover network and matched perfectly to the other drivers. That change is an 8-ohm resistor across the tweeter. Normally the impedance of the tweeter would be above 200 ohms at 6000 Hz, so the crossover network would not be effective for it. But by connecting an 8-ohm resistor in parallel with the tweeter, the combination 3.3 μf capacitor in the crossover network and the resistor gives a roll-off of low frequency energy in the tweeter circuit that becomes effective at 6000 Hz. The crossover circuit has built-in L-pads that provide level controls for the mid-range driver and the tweeter.

Building the Box

You can build the basic box from 3/4″ particle board, and then cover it with any kind of veneer (Fig. 5-1 and Table 5-1). Here 1/4″ birch plywood was chosen for the exterior surfaces, so the procedure will be slanted toward using plywood as a veneer. If you wish, you can alter the construction to match that of any other suitable material, but keep to the same internal dimensions. As an example, you can make the outer walls of solid pine and reinforce them with 1/2″ particle board, as was done in Project 6. In that case you would

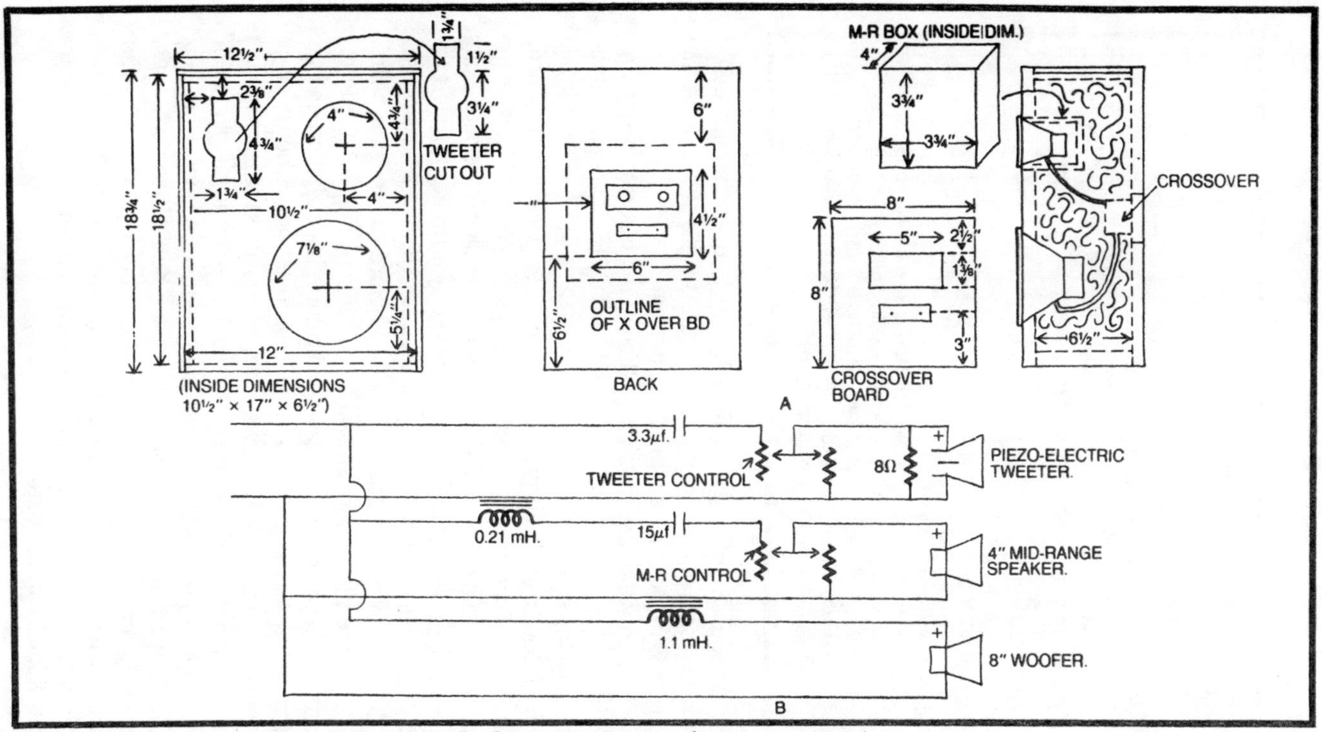

Fig. 5-1. (A). Construction plans for Project 8. (B). Schematic diagram of crossover network.

Table 5-1. Materials list for Project 8.

¾″ Particle board for main box, pieces needed:		
2	6-½″ × 12″	Top & bottom
2	6-½″ × 17″	Sides
2	12″ × 18-½″	Speaker board & back.
¼″ Birch plywood for veneer, pieces needed:		
1	12-½″ × 8″	Top, dadoed on ends.
2	18⅝″ × 8″	Sides.
½″ plywood for mid-range box, pieces needed:		
2	4″ × 4-¾″	Top & bottom
2	4″ × 3-¾″	Sides
1	4-¾″ × 4-¾″	Back
½″ plywood for grille frame, optional:		
1	12-½″ × 18-¾″	
Speakers and components:		
1	8″ high compliance woofer, Radio Shack #40-1006.	
1	4″ high compliance speaker, Radio Shack #40-1197.	
1	2″ × 5″ piezo-electric tweeter, Radio Shack #40-1280.	
1	3-way crossover, Radio Shack #40-1339A (1000 & 6000 Hz).	
1	8 ohm resistor, 1 watt or higher.	

have to add about ½″ to the width of the top, bottom, and sides. Add 1″ to the length of the top, add 1-¾″ to the length of the sides, and subtract ½″ from the length of the bottom.

To follow the construction method of the box pictured in Fig. 5-2, start by cutting out the particle board walls. Glue and nail the bottom to the sides and then the top. Cut the 4-½″ × 6″ window in the back before you glue and nail it to the sides. Typical boxes made by this kind of construction should have the front installed before the back so the front corners can be caulked. This allows the back joints to be caulked after the front is installed by working through the speaker holes, but in this case it is necessary to put the back on first so the crossover board can be installed before the speaker board.

Prepare the crossover board according to Fig. 5-1A. Set the box on its back and run a bead of silicone rubber glue around the cutout in the back and press the crossover board down on it. Remove the four small screws from the cover on the crossover network and take off the cover. Scrape, file, or grind off a ¼″ to ½″ strip of paint around the edge of the crossover's front panel (Fig. 5-3). This roughened metal will permit a better bond when you mount the crossover to the board. Run a bead of silicone rubber around this prepared border and press the crossover network down over the crossover board in the back of the enclosure, centering the control knobs in the cutout.

Fig. 5-2. Enclosure for Project 8 is made of ¾″ particle board with a veneer of ¼″ birch plywood.

Prepare the speaker board by cutting out the holes for the speakers and installing the mid-range box to it. When you cut out the parts for the mid-range box, drill a ¼″ hole in the center of the back piece to admit the speaker cord. Glue and nail together the sides and back of the mid-range box. Place the speaker board on a

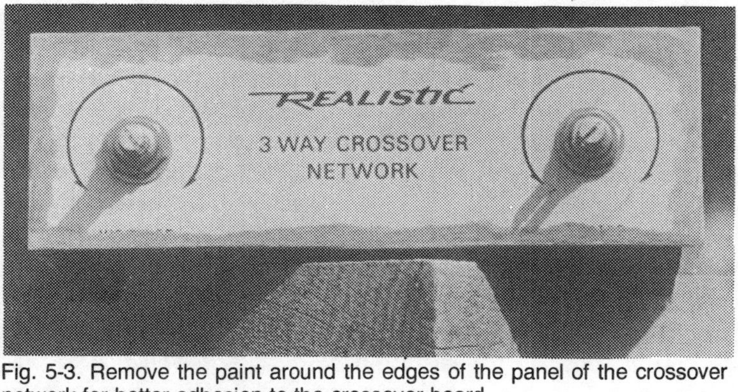

Fig. 5-3. Remove the paint around the edges of the panel of the crossover network for better adhesion to the crossover board.

Fig. 5-4. Use an old blanket or other protection while sanding the birch plywood exterior.

firm support and start #6-penny finishing nails through the speaker board. These should be spaced about 5″ apart and located on lines ⅜″ from each edge of the board. Drive in these nails until they barely penetrate the lower surface of the board. Use glue and smaller #4 finishing nails to attach the mid-range box to the speaker board. Start the nails in the same way as the larger nails; then glue the front edges of the mid-range box and drive down the nails.

Caulk the interior joints of the larger enclosure and those of the mid-range box. Cut three pieces of lamp cord about 24″ long for speaker leads. Examine the lamp cord to identify the ribbed lead on each cord. Connect the cords to the proper terminals on the rear of the crossover network. Cut a short piece of lamp cord and use it to connect the main box terminals to the input terminals of the crossover network. Line the interior surface of the main box with damping material and put a layer on the lower side of the mid-range box.

Examine the contact surfaces of the speaker board and the front edges of the main box. If these surfaces are smooth, you can depend on wood glue to seal the box. If the fit is uneven, use silicone rubber glue. Coat both matching surfaces with glue. Then feed the speaker leads to their proper route, the tweeter lead out through the tweeter cutout and the mid-range lead through the ¼″ hole in the back of the mid-range box as you place the speaker board down on the box. It's a good idea to have someone help with this step to keep the speaker cords from getting into the glue on the edge of the box. Drive down the nails that attach the speaker board to the enclosure walls.

Use a wood rasp or sandpaper to remove any protrusions, such as top overhang, from the box. Cut out the birch plywood parts and fit them to the box. Although the sides are listed at 18-¾″ long, you may have to shorten these parts, depending on whether you dado the edges of the top plywood or leave the edges a full ¼″ thickness. If you want a minimum of end grain to show, you should remove a ¼″ wide × ⅛″ deep strip from the lower surface at each side of the plywood top piece. Attach the plywood to the particle board box with glue or panel adhesive. When the adhesive sets, sand the plywood and stain it (Fig. 5-4). Apply the final coats, either varnish or, if you prefer, a wax or oiled finish.

Wiring

Connect an 8-ohm resistor across the terminals of the tweeter before you install it (Fig. 5-5). Set the tweeter on the speaker board

Fig. 5-5. Don't forget to wire an 8-ohm resistor across the terminals of the piezo-electric tweeter to match it to the crossover network.

to connect and solder the lamp cord leads. On this and other speaker connections make sure the polarity is right; just follow the directions with the crossover network. Mount the tweeter with a gasket of silicone rubber, but use screws to insure firm installation. Silicone rubber does not adhere well to smooth plastic surfaces.

When you wire the mid-range driver and the woofer, you should be alert that the right terminal is the positive one on the mid-range speaker, the left one on the woofer. Before you install the mid-range speaker, fill the mid-range compartment with loose polyester fiber or similar damping material. There is no need to use screws for the mid-range speaker or the woofer if you use silicone rubber glue to install them. Just press the speakers firmly down on a continuous bead of silicone rubber, and leave the cabinet on its back until the compound sets, at least a half day or preferably for a full day.

That's it except for a grille frame and your choice of grille. As in all speaker systems you should choose an open grille, one that won't muffle the upper frequencies of the super tweeter. If you place the speaker in a horizontal position, as on a bookshelf, don't be concerned whether the tweeter is above or below the mid-range driver. Just make sure that nothing blocks the sound of the tweeter. A pair of these will give very satisfying stereo sound, especially considering the size of the systems.

PROJECT 9: THE TRADITIONAL

- **Construction:** Use power saw.
- **Wiring:** Fairly simple. Commercial crossover.
- **Comments:** Competent sound. Strong bass. Good power handling.

This kind of speaker, a 3-way system based on a 10″ acoustic suspension woofer in a compact closed box, has probably been used in more good quality speaker systems than any other kind since the advent of stereophonic sound (Fig. 5-6). Such a combination is no guarantee of high fidelity, of course; each system must stand on its own. This one does that very well.

The "D" grade of white pine used in this project is about as expensive as any ordinary building material, so if you like other materials better than pine, you should make a substitution. If you change to plywood, you may want to alter the plans so that less end grain shows on the edges of the top. Another possibility is to use a lower grade of pine. It will have a few knots and other imperfec-

Fig 5-6. Project 9 completed.

tions, but these blemishes won't affect the sound. The particle board liner assures you of adequate strength.

Note that the enclosure shown in Fig. 5-6 was made with cleats at the back to receive screws that hold the back on. This was necessary in the prototype of this project and the next one so that various crossover networks could be tested. If you want to install the back permanently, you can simplify construction slightly. Just eliminate the dado cut for the rear cleats, extend the particle board liner back ¾″ farther, and glue in the back.

Cutting The Pine Parts

Begin by cutting the pine parts to size. If you buy the particle board in a large sheet, you should rough cut it to approximate size, then trim each piece to the size you need. This gives you easy-to-handle pieces for the cuts that must be accurate. It's a good idea to postpone the final cut on the particle board pieces until you are ready to use them.

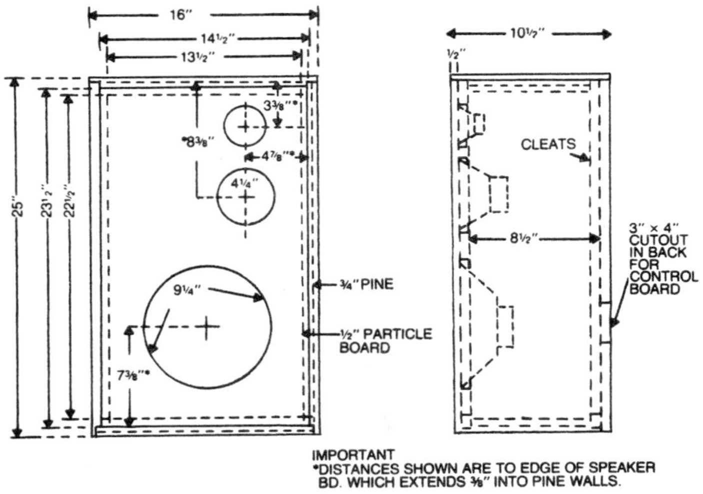

Fig. 5-7. Construction plans for Project 9.

Cut the rabbet on the top and sides of the pine. See Fig. 5-7 and Table 5-2. Note that these cuts must make a space for a ¾" board, so the cuts will each be ¾" wide by ⅜" deep. Next make the dado cuts for the grooves in the top, bottom and sides (Fig. 5-8). The front groove should be recessed ½" from the front edge of each board, and the rear groove ¾" from the back edge. The grooves should be ¾" wide by ⅜" deep.

Set up the pine parts to see if they fit together as planned. Then cut out the speaker board. Note that this board must be about ¾" larger in each dimension than the opening at the front of the enclosure. Cut the speaker holes in the board. Set up the bottom, a side, and the speaker board to see how they fit together as shown in Fig. 5-9. Note that the speaker board should reach to the top edge of the side when it is set in the groove on the bottom.

Assembly

Begin assembly by gluing and putting together the bottom, side and speaker board. Use glue sparingly to avoid staining the exterior surface of the boards. Glue and attach the other side and the top. Clamp the box together with bar clamps. If you have no bar

Table 5-2. Enclosure materials for Project 9.

¾" Pine boards, pieces needed:		
2	10-½" × 24-⅝"	Sides
1	10-½" × 16"	Top
1	10-½" × 15-¼"	Bottom
¾" Particle board, pieces needed:		
1	15-¼" × 24-¼"	Speaker board
1	14-½" × 23-½"	Back
½" Particle board, pieces needed:		
1	7-¾" × 14-½"	Top (Install First)
1	7-¾" × 23"	Sides
1	7-¾" × 13-½"	Bottom (Install Last)
¾" × 1-⅛" pine cleats, pieces needed:		
2	24-¼" long	Side cleats for back
2	Approx. 13" long	Top & bottom cleats for back.
⅜" plywood for grille frame, optional.		
1	14-7/16" × 23-7/16"	Grille frame with cut-outs to fit over speaker frames.

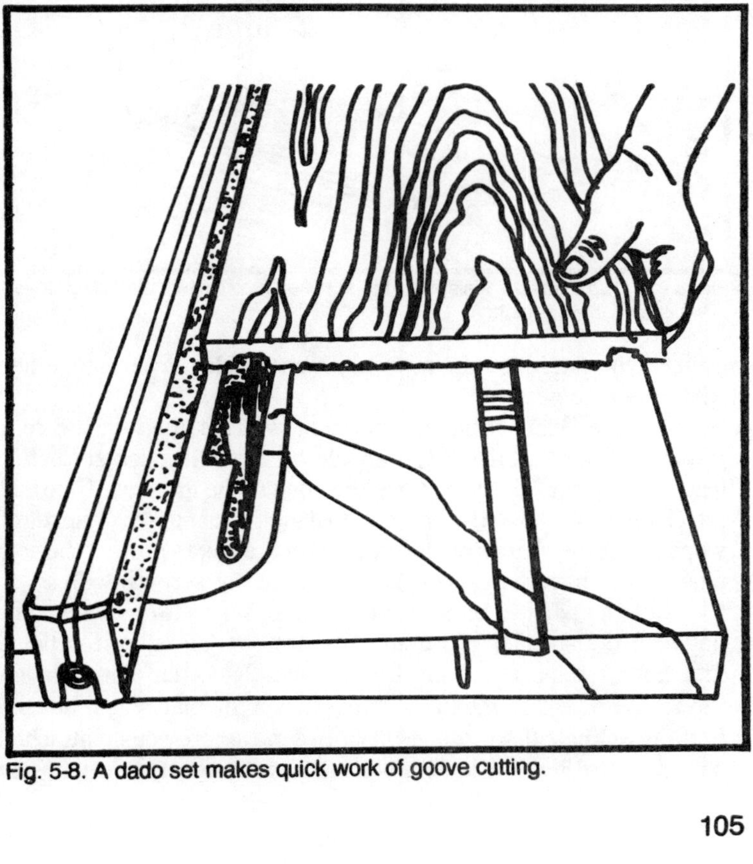

Fig. 5-8. A dado set makes quick work of goove cutting.

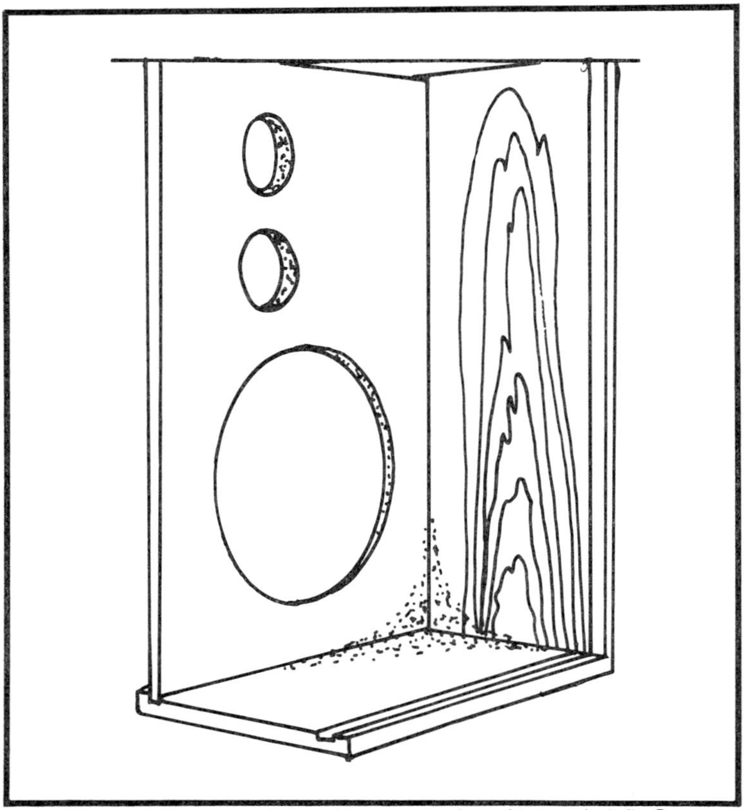

Fig. 5-9. Set up a bottom, a side and the speaker board to test their fit. Grooves in walls lock them together.

clamps, you can substitute rope or straps to hold the parts together while the glue sets.

Cut the cleats to fit the space allowed for them in the rear grooves. Note that the cleats should be about 1″ thick so that at least ⅝″ of cleat will show above the edge of the grooves. Glue and attach the cleats. Use C-clamps to hold them, but put small pieces of wood under the clamps on the exterior of the box to protect the soft pine. If you have no C-clamps, use #8×1-¼″ screws.

Cut the particle board liner to the exact width needed to fit between the speaker board and the cleats. Glue in the top piece first, then sides and bottom. By following this order of installation the top piece will fit the full width between the pine sides, but the bottom will be cut shorter to fit between the previously attached particle board liners. Use scraps of wood wedged between opposing sides to apply pressure to the particle board liner while the glue sets (Fig. 5-10).

Cut out the back to fit the space between the pine walls. Drill shank holes for screws at 5″ intervals around the back. Make a 3″ × 4″ cutout for the crossover board, centered in the back 7″ above the bottom. This will place the crossover network directly behind the position of the woofer in the front. If there is any need in the future to make changes in the wiring, it can be reached by removing the woofer and working through the speaker hole.

Cut a 6″ × 8′ piece of ¼″ tempered hardboard to make the crossover board. Drill two ⅜″ holes for the mid-range and tweeter controls, spaced 2″ apart. Then locate the holes for a screw terminal strip on a line 1-¼″ below the holes for the controls. Install the terminal strip on the smooth side of the hardboard with glue and tacks. Solder two 4″ lengths of wire to the terminals, one to each terminal. Install two 8-ohm L-pads in the previously drilled holes (Fig. 5-11). If you intend to use this speaker system with high-powered amplifiers and drive it hard, you should install heavy duty L-pads, at least for the mid-range control.

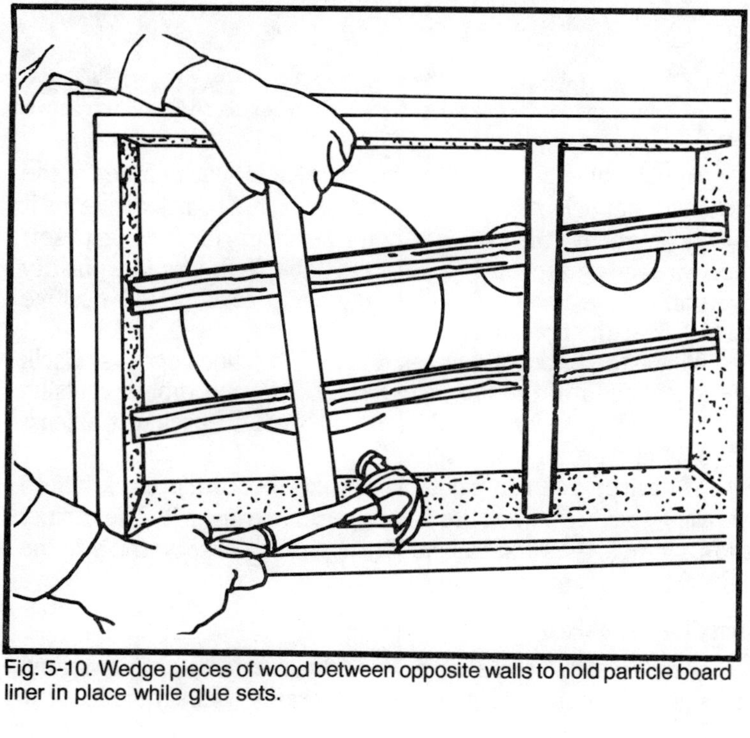

Fig. 5-10. Wedge pieces of wood between opposite walls to hold particle board liner in place while glue sets.

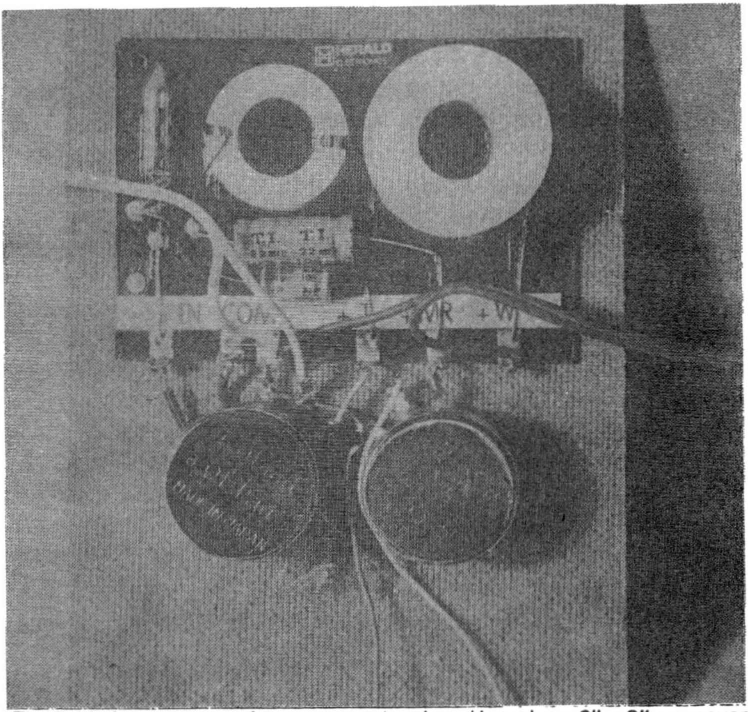

Fig. 5-11. Arrangement of crossover network and L-pads on 6" × 8" crossover board.

Apply a blob of silicone rubber glue under each corner of the crossover network card and stick it down on the board, centering it in the space above the L-pads. Allow a few hours for the glue to set; then wire the crossover (Fig. 5-12 and Table 5-3). Observe polarity by running the beaded lead of your lamp cord to the positive terminals at the crossover and speakers.

Install the back with glue and #8 × 1-¼" wood screws. Caulk the inner joints of the box with latex or silicone rubber caulking compound. Line the walls of the box and cover the back with at least 2" to 3" of damping material.

If desired you can paint the speaker board flat black before installing speakers. Sand the pine exterior and finish it. The cabinet shown in Fig. 5-13 was finished by applying a clear varnish to the light colored pine.

Installing Speakers

Next, install the speakers. Feed the cord for each speaker out through the appropriate hole and temporarily connect the leads to

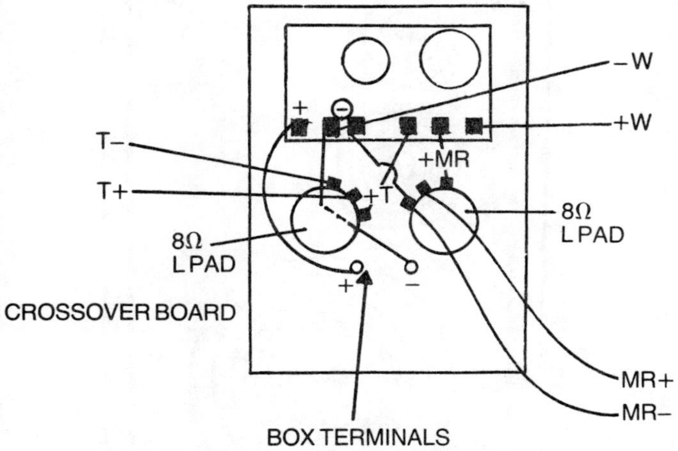

Fig. 5-12. Crossover wiring for Project 9.

the speakers. Set the speakers in their mounting holes. Turn off the power to a receiver or amplifier and turn the volume all the way down. Connect the receiver to your speaker. You will have to prop the cabinet across an empty carton or use some other kind of prop to do this. Turn on the receiver and turn the volume up so you can hear the speakers easily. Listen carefully to each speaker to see if the bass is coming from the woofer and the upper highs from the tweeter. Adjust the tweeter and mid-range controls to see if they are operating properly. When you are satisfied that the wiring is right, turn off the receiver and disconnect the speaker. Solder the connections to each speaker. Run a bead of silicone rubber glue around each speaker hole and install the speakers by pressing each one down on the glue. Rotate it slightly to spread the glue and then

Table 5-3. Speakers and components needed for Project 9.

1	10″ woofer, Pioneer model A25FU10-OF
1	4-½″ mid-range speaker, Pioneer A11EC65-01F
1	1-¾″ tweeter, Pioneer B45BP50-01F
	Speakers available from McGee Radio Co.
1	3-way crossover network, Herald CN31A/P(OR CN31A) with crossover frequencies at 700 and 4500 Hz.
2	8 ohm L-pads.

Fig. 5-13. Project 9 and drivers used in the system.

press down hard on the speaker frame. Leave the cabinet on its back for several hours for the glue to set. Don't drive the speaker hard for at least a day.

Cut a grille frame from ⅜" plywood or particle board to fit inside the pine walls of the cabinet at the front. Make the speaker cutouts, and paint the grille frame flat black. Stretch a grille cloth over the board and staple it on the back. Install the board on the cabinet. This completes your speaker.

PROJECT 10: THE HEAVYWEIGHT

- **Construction:** Use power saw.
- **Wiring:** Homemade crossover. Can substitute commercial model.
- **Comments:** Good efficiency for acoustic suspension speaker. Smooth well-defined bass.

The name "Heavyweight" was suggested by the magnet on the woofer used in this project, a 38-ounce slug of iron. This heavier than average magnet gives greater efficiency and better damping on the woofer than is typical of speakers designed for closed box operation.

The homemade crossover network specified for this speaker has some unusual values in it, chosen by listening tests rather than from formulas. The first crossover tried was the same commercial model used in Project 9 (Fig. 5-14). The performance with that network was satisfactory, but tests indicated that some improvement could be achieved by adding impedance equalizers across the woofer and mid-range speakers. Then a response shaping network was added until there were 16 elements in the network. The next step was to simplify the network. The final choice is shown in Fig. 5-15. It gave performance that sounded very similar to that obtained by the much more complicated crossover. The choke in series with the woofer is larger than expected, but it produces a more effective

Fig. 5-14. Commercial crossover network shown here was later abandoned for the one in Fig. 5-15.

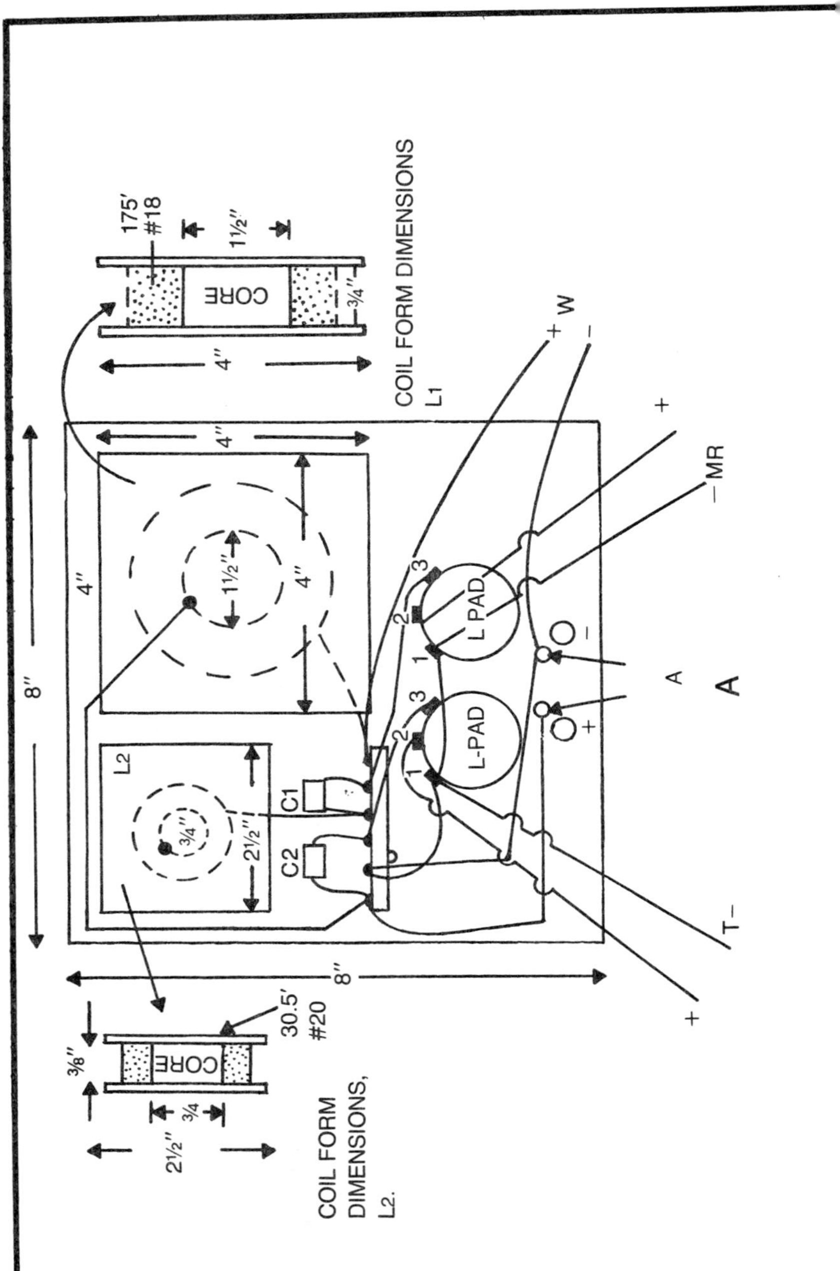
COIL FORM DIMENSIONS
L1
175'
#18
1½"
CORE
3/4"
4"
4"
4"
8"
1½"
4"
L PAD
L-PAD
1
2
3
+ W
−
+
MR
−
A
A
+
−
C1
C2
L2
2½"
3/4"
T−
+
8"
30.5'
#20
CORE
3/8"
3/4
2½"
COIL FORM
DIMENSIONS,
L2.

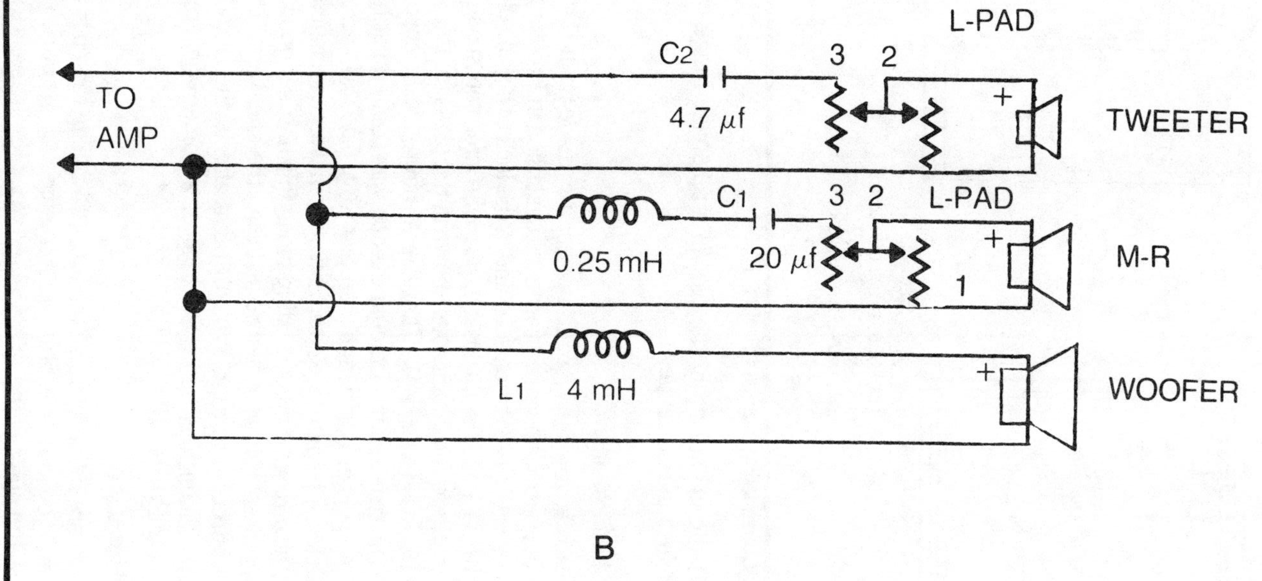

Fig. 5-15. (A). Suggested crossover board plan. (B). Circuit for homemade crossover board.

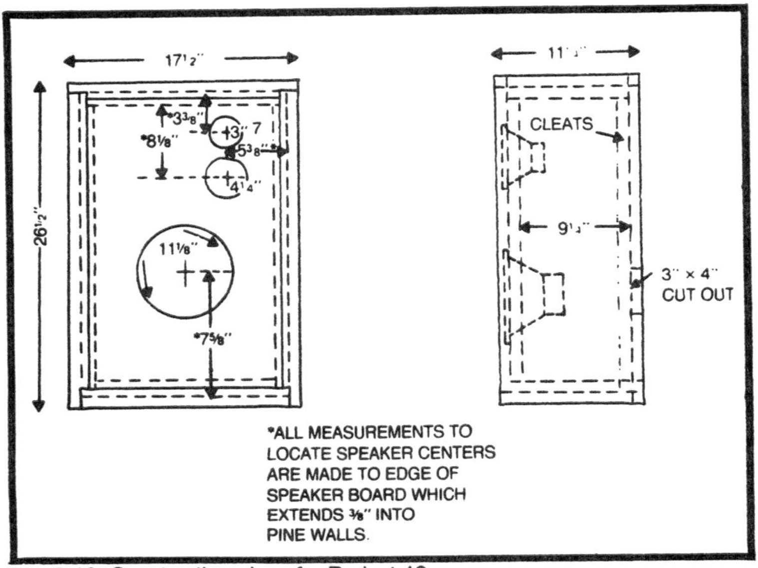

Fig. 5-16. Construction plans for Project 10.

roll-off at the upper end of the woofer's response range with no apparent loss of mid-range.

Construction Procedure

The enclosure used for this project was made from "D" grade of white pine with the same procedure described in Project 9. The differences are in the dimensions and in the lack of a grille cloth for Project 10. To build the enclosure, go to the information on Project 9, but substitute the dimensions shown in the plans for this project (Fig. 5-16 and Table 5-4). There is a difference in appearance because the front edge of the pine sides was painted flat black to match the speaker board in this project (Fig. 5-17). No grille cloth was installed. The pine sides, top and bottom were varnished with a clear varnish to give a contrasting blond finish. It's a good idea to apply any finish you want before installing the speakers. The only change of procedure from Project 9 that affects the sound is the change of dimensions and the wiring of a homemade crossover network for Project 10.

Wiring Procedure

If you are intimidated by the idea of winding the coils for the crossover, you can substitute the commercial network used in Project 9. But don't let the coils frighten you; they are easy to make.

For a stereo pair of these speakers you will need 2 pounds of #18 gauge magnet wire and a half pound of #20 gauge wire. To make the core for the larger coil you can cut a 1-½″ disc from a ¾″ piece of particle board or plywood. For the smaller core a piece of ¾″ wood dowel is the obvious choice, although you can use a section of any plastic container or cardboard tube that is ¾″ in diameter (Fig. 5-15A).

Make up the coil forms by gluing hardboard or thin plywood sides, or cheeks, to the core. Or you can bolt the pieces together to make a temporary form. Then remove the form and wrap the coil with tape. Just make sure you have no metal in the core of the coil when it is used.

Drill a hole in one side of each coil form, located next to the core. To wind the coil, pass the end of your magnet wire through the hole from inside the form to the outside. Leave about 6″ protruding and start winding. If you carefully measure the length of the wire, you do not need to count the turns. For the larger coil you will need 175′ of #18 gauge wire per coil, but allow 12″ extra for two 6″ end leads. The smaller coils require 30-½′ of wire plus 12″ per coil. The

Table 5-4. Materials list for Project 10

1″ × 12″ pine boards, pieces needed: (Actually ¾″ × 11-¼″.)		
2	11-¼″ × 26-⅛″	Sides
1	11-¼″ × 17-½″	Top
1	11-¼″ × 16-¾″	Bottom
¾″ Particle board, pieces needed:		
1	16-¾″ × 25-¾″	Speaker board
1	16″ × 25″	Back
½″ Particle board, pieces needed:		
1	7-¾″ × 16″	Top, install first.
2	7-¾″ × 24-¼″	Sides
1	7-¾″ × 15″	Bottom, install last.
¾″ × 1-⅛″ pine cleats.		
2	25-¾″ long	Sides
2	14-½″ long	Top & bottom
Speakers and components list:		
1	12″ woofer, CTS 12W38C	
1	4-½″ mid-range speaker, Pioneer A11EC65-01F	
1	1-¾″ tweeter, CTS 2TA8	
	Speakers available from McGee Radio Co.	
2	8 ohm L-pads	
1	4.7μf non-polarized electrolytic capacitor.	
1	20μf non-polarized electrolytic capacitor.	
1 lb. #18 magnet wire		
1/10 lb. #20 magnet wire		

easiest way to make the coils is to measure the correct length of wire, cut it from the spool and wind the coil. You don't have to count the turns, but the large coils will require about 280 turns. The smaller ones take 100 turns. If you have access to a bridge, you can add several feet of wire to the length. Then measure the inductance of your coil with the bridge. Remove enough turns to obtain the precise value of inductance specified.

When you finish winding your coil, tape the wire into the form so it can't loosen. Remove the insulation from the ends of the coil by scraping, filing, or heating the wire with a soldering iron or gun. Some kinds of magnet wire insulation are easy to remove with a hot iron; others require scraping.

The capacitor in series with the tweeter is a standard value, available at most electronics stores. For the 20 μf mid-range capacitor you will probably have to put two capacitors in parallel, such as a 16 μf and a 4 μf, or a 12 μf and an 8 μf capacitor.

Crossover Board

When you have assembled the coils and the capacitors, cut an 8″ × 8″ piece of ¼″ tempered hardboard for a crossover board. Drill two ⅜″ holes for L-pads at a distance of 2″ center to center from each other, placed as shown in Fig. 5-15A. Then locate the holes for the screw terminal strip on a line 1-¼″ below the holes for the pads. Put the terminal strip on the smooth side of the board with glue and tacks. Install the L-pads with their hardware. Then mount the two coils and a 6-lug terminal strip, as shown in Fig. 5-14, with Elmer's Heavy Grip Cement or Leech's Heavy Adhesive. Follow the Fig. 5-15B and Fig. 5-14 to wire the crossover network, including speaker leads at least 24″ long. Place the crossover network over the cutout on the back with silicone rubber glue.

Installing the Back and Speakers

Test the wiring before permanently installing the back and the speakers. Set the cabinet on its back, propped up so you can reach the terminals and controls on the back. Temporarily connect the leads to each speaker and set them in the holes. Turn off your receiver and turn the volume all the way down. Connect the receiver to the speaker terminals on the back and turn it on. Bring the volume up just enough to hear the sound easily. Check to see that the bass comes from the woofer, the middle tones from the mid-range speaker, and the highs from the tweeter. Operate the controls and see if they work properly. Then disconnect the receiver

Fig. 5-17. Instead of a grille cloth for Project 10, the front edge of the pine walls were painted black to match the speaker board.

and solder the speaker connections. Then, after caulking the inner joints and adding 2″ to 3″ insulation on the enclosure walls and back, install the back with glue and screws. Install the speakers with silicone rubber glue.

You may have noticed that the mid-range speaker specified for this project is the same one used in Project 9. Another mid-range speaker is shown in the picture of Project 10 with an extra set of drivers on the table. After trying several of similar cost, the Pioneer model listed in Table 5-4 was considered the best choice.

PROJECT 11: THE TRI/LINEAR

- **Construction**: Easy. But can use power saw on plywood veneer.
- **Wiring**: Commercial crossover with minor changes.
- **Comments:** Warm bass. Smooth mid-range and treble.

Fig. 5-18. Free-standing Project 11 occupies little floor space.

This floor model speaker has a rare combination of virtues: easy construction and an elegant appearance (Fig. 5-18). The latter, of course, depends on how you handle the plywood veneer. For a refined look you can put on a stretch fabric grille cloth, but the brushed aluminum tweeters and mid-range speakers beg to be seen. If you add a grille, choose carefully. You won't want to muffle the extended range of this super tweeter. These tweeters are sold in small enclosures for external installation as add-on drivers to extend the range of full speaker systems. Here they are removed from their cases and mounted in the cabinet. The crossover components from the tweeter are transferred into the crossover network. This combination gives a sharper cutoff rate for the high channel than was designed into the crossover network, insuring good protection for the tweeter.

If you are partial to large floor speakers and like an airy high end response, this, the largest closed box speaker in the book, may be for you. Acoustic suspension woofers are normally installed in smaller enclosures, but the optimum enclosure depends on several factors. Here, instead of squeezing the woofer into the smallest box suitable for satisfactory performance, it is coupled to an enclosure designed to allow its most extended bass response. A larger box would probably make the bass weak and over damped. Smaller boxes would yield more prominent bass but with higher frequency of system resonance. When tests indicated that the woofer would

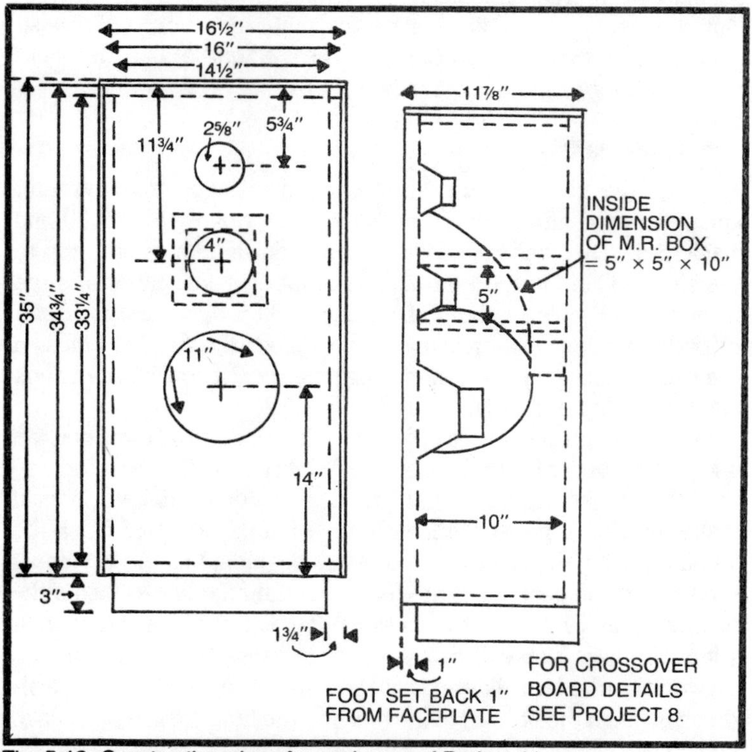

Fig. 5-19. Construction plans for enclosure of Project 11.

Table 5-5. Materials list for enclosure of Project 11.

Qty	Size	Use
¾″ Particle board, pieces needed:		
2	10″ × 33-¼″	
2	10″ × 16″	
2	16″ × 34-¾″	
⅜″ Plywood, one piece needed:		
1	16″ × 34-¾″	Faceplate
¼″ Birch plywood, pieces needed:		
1	11-⅞″ × 16-½″	Top
2	11⅞″ × 34-⅞″	Sides
¾″ material for foot, pieces needed:		
1	3″ × 13″	Front piece of foot
2	3″ × 10-⅞″	Side pieces for foot.

¾″ P.B for mid-range Box	Qty	Size
Sides	2	6½″ × 10″
Top & bottom	2	5″ × 10″
Speaker board and back		

perform well in a volume of 3 cubic feet, that volume was chosen. In addition to optimum bass range, it gives the visual interest that a tall floor model offers. There is a theoretical disadvantage with the shape. One internal dimension is a bit long compared to the others. But the interior is broken up by the mid-range box, which also serves the function of bracing the speaker board and the back, making a more rigid enclosure.

Cutting the Parts

Cut the parts for the basic enclosure, sides, top, bottom, speaker board and back, from ¾″ particle board (Fig. 5-19 and Table 5-5). Glue and nail, using 6-penny finishing nails, the top to the sides. Then do the bottom. Make the speaker cutouts in the front panel. Prepare a sheet of ⅜″ plywood for the faceplate. Set it over the speaker board and mark an outline of the speaker holes on the back of the faceplate. Then install the speaker board on the box shell with glue and nails.

Next, cut three pieces of ¾″ material to make the foot. You can use particle board for the foot and paint it black or, if you prefer, use pine or plywood. If you use hardwood plywood, as I did, you should make beveled joints at the front corners of the foot and finish the wood to match the sides and top. Mark an outline on the bottom of the box for the position of the foot, indenting the line ⅝″ from the front edge and 1-½″ from each side edge. Cut the foot pieces to fit within this line. These distances will be increased when the exterior plywood is installed. Glue and nail them to the box, driving your nails down through the bottom of the box into the foot. It's a good idea to start these nails with the box sitting on the floor of your workshop to support the particle board bottom, until the nails are driven in far enough to just penetrate the lower surface.

Cut out the parts for the mid-range box. Glue and nail the shell of the tube together. Check the length of the tube; it should be exactly the same as the interior depth of the large enclosure so it will contact the back. Place the large enclosure on its face and set the mid-range tube in position (Fig. 5-20). Mark an outline of the mid-range box on the inside of the speaker board. Cut three small ¾″ × ¾″ cleats to fit around the mid-range box on the back of the speaker board. Install these cleats with nails and glue at the marked position of the mid-range box so that they will extend around it on the sides and the top. No cleat should be installed on the lower side of the mid-range box, next to the woofer.

Before installing the mid-range box, drill a ¼″ hole through the bottom about midway between the front end and back end of the

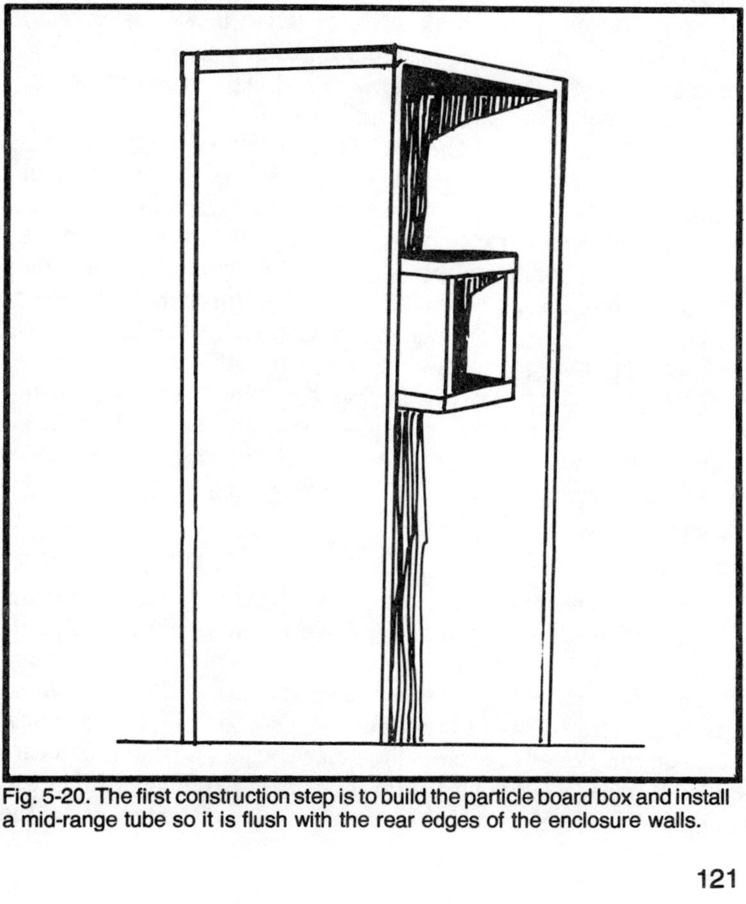

Fig. 5-20. The first construction step is to build the particle board box and install a mid-range tube so it is flush with the rear edges of the enclosure walls.

tube. Glue the front edge of the mid-range box and set it upright on the floor. Place the large enclosure over it, using the cleats on the lower side of the speaker board to guide the parts together. Nail the speaker board to the mid-range box. Note that if the box is in the proper position, the front will be supported by it. Nails should go right through the hard particle board without bending.

Working on the Enclosure

Place the back on the floor and set the enclosure over it. Reach through the woofer hole and mark the outline of the mid-range box on the back. You will be able to easily reach the bottom and side of the tube. You will locate the top line by measure. Remove the enclosure and drill small guide holes through the back to properly position the nails that will attach the back to the mid-range shell. Note that these holes should be located ⅜″ inside the outline that you have marked on the back. Make a cutout in the lower back, centered behind the woofer position, for the crossover board. Set the back face down on a firm surface and start nails through it, spaced at about 5″ intervals and indented ⅜″ from the edges.

Before installing the back, caulk the inner joints of the large enclosure and caulk around the mid-range box on the speaker board. Cut a 30″ length of lamp cord and feed it through the hole in the bottom of the mid-range tube. Allow the cord to extend out the front of the speaker board by at least 6″. Then caulk the hole in the tube from both sides. Put another 30″ length through the tweeter hole and around one side of the mid-range box. Secure this cord by tape to avoid losing it in the next steps. Set the enclosure on its face and cover the surface in the upper part of the enclosure with damping material. You can drape a horseshoe shaped roll of thick batting over the mid-range tube. Cover the upper half of the back with damping material. Glue the matching surfaces and nail down the back.

Faceplate

Now get the previously marked faceplate and locate the center of each speaker position that you have marked on the back of it. Drive a small nail through the center of the woofer position and use this nail as a guide for your saber saw to make the 12-⅛″ cutout so the woofer frame will fit inside the ⅜″ plywood. Drill a small hole through the center of each of the other speaker positions to locate those cutouts. At the tweeter position draw two concentric circles, one 2-⅝″ in diameter and the other one 4″ in diameter. Make the

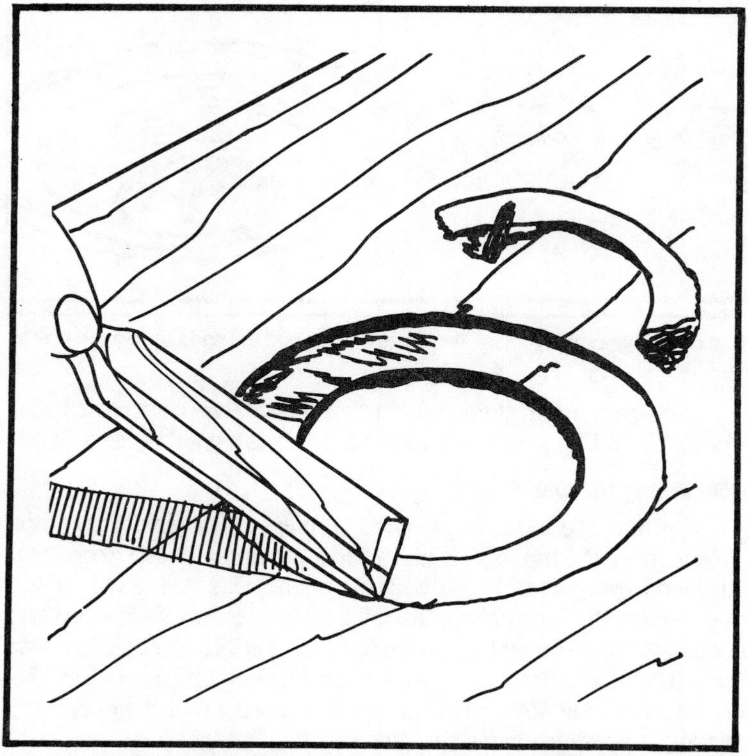

Fig. 5-21. Use a wood chisel to remove enough material to permit the tweeter to set flush with the front surface of the faceplate.

2-⅝″ cutout. For the mid-range speaker, draw lines at the approximate location of the speaker cutout line. These lines should be horizontal and vertical, each located at 2-⅝″ from the center hole. Then center the mid-range speaker in these lines, face down, and mark a heavy line around the edge of it to guide your saw for the cut.

To flush mount the tweeter, it is necessary to remove about ⅛″ of plywood in a ring between the tweeter cutout and the 4″ line marked earlier. You can do this by careful work with a wood chisel (Fig. 5-21).

Set the enclosure on its back. Use 4-penny finishing nails to temporarily nail the faceplate to the enclosure. Try the speakers to see if they fit perfectly (Fig. 5-22). If necessary, use a wood rasp or sandpaper to enlarge any of the cutouts in the faceplate so the speakers fit flush. When you are satisfied with the fit, remove the speakers; then glue and nail down the faceplate.

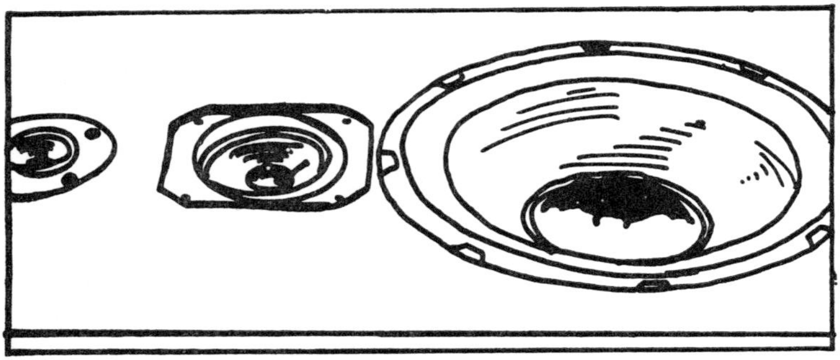

Fig. 5-22. Temporarily install faceplate on box and set speakers in the box to test the fit.

You can either stain and varnish the faceplate or paint it flat black (Fig. 5-23). If you plan to paint it, this is a good time to do so.

Completing the Cabinet

Cut out the exterior plywood veneer, starting with the top. After cutting the top to size, use a dado or your saw blade to make a cut along each side ¼″ wide × ⅛″ deep. This will leave an ⅛″ overhang at the top corners. Install the top piece with glue or panel adhesive. Coat the matching surfaces and set the top upside down on a firm surface covered by an old blanket or towel. Set the enclosure upside down over the top so the weight of the enclosure will apply pressure while the glue sets.

Cut the side pieces to fit the same way you attached the top. When the glue has set, sand the cabinet and apply stain and a final finish as described in Chapter 1.

CROSSOVER MODIFICATION

Prepare the crossover board according to Fig 5-19. Install a speaker screw terminal strip with glue and tacks. Run a bead of silicone rubber around the edge of the cutout in the back. Pass the crossover board through the woofer hole and press it down on the back, centering the speaker terminals and the cutout for the crossover panel in the back cutout.

Check to see that you have the 40-1339A crossover network; the letter A must be present in the catalog number. Remove the four small screws from the crossover network cover and take the cover off. File, scrape, or grind off a strip of paint around the front panel of the crossover. This strip of roughened bare metal, about ¼″ to ½″ wide, will permit better bonding between the crossover network panel and the crossover board.

Crossover Network Changes

It is necessary to make some changes in the crossover network before installing it in the box. First remove the tweeter from its small enclosure by removing the screws that hold the tweeter and enclosure together. Carefully remove the push-on connectors to the tweeter terminals. Next, remove the tweeter crossover components from the shell. A 5.5 mm socket is ideal for holding the nut inside the shell while you unscrew the bolt in the back, but you can probably use pliers. The crossover components that you remove, a capacitor and a choke, should be installed in the main crossover network, as follows (Fig. 5-24 and Table 5-6).

Locate the orange wire that goes to the T+ terminal. Cut this wire loose from the terminal. Note that this wire goes to the middle lug on the tweeter control. Cut the wire loose at the control. Connect the free end of the 2.7 μf capacitor, the end with the red

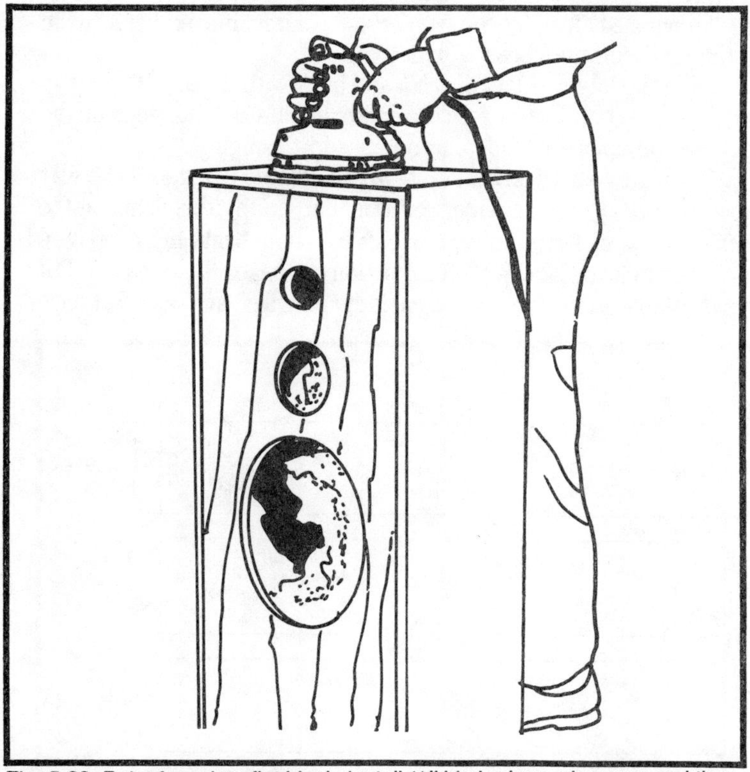

Fig. 5-23. Paint faceplate flat black, install ¼″ birch plywood veneer and then sand and finish the exterior.

spaghetti on it, to the middle lug on the tweeter control. Join the other end of the capacitor, the end connected to the choke, to the T+ lug on the tweeter terminals. Connect the free end of the choke to the T-lug on the tweeter terminals. Solder these connections.

Run a bead of silicone rubber glue around the bare metal border of the crossover panel. Press the crossover network down on the hardboard in the back of the speaker enclosure. Leave the parts in position for several hours before disturbing them.

Wiring the Network

Next, wire the crossover network. Run a short piece of lamp cord from the terminals on the board to the input of the crossover network. Locate the cord from the mid-range box and connect it to the mid-range terminals on the crossover network. Observe polarity in each hookup. Locate the cord from the tweeter hole and connect it to the tweeter terminals on the crossover network. Finally, cut a 24″ lead for the woofer and connect one end of it to the W terminals on the crossover network.

Line the lower sides, bottom and back with at least 2″ to 3″ of damping material. Cut a slit in the damping material that goes on the back so the speaker leads can be passed through it.

Turn off your receiver and turn the volume control all the way down. Connect your receiver, or amplifier, to the terminals on the back of the enclosure. Lay the enclosure on its back, but prop it up on some bricks or blocks of wood so you can reach the controls. Put the speakers on the face of the cabinet and wire each one. Set each

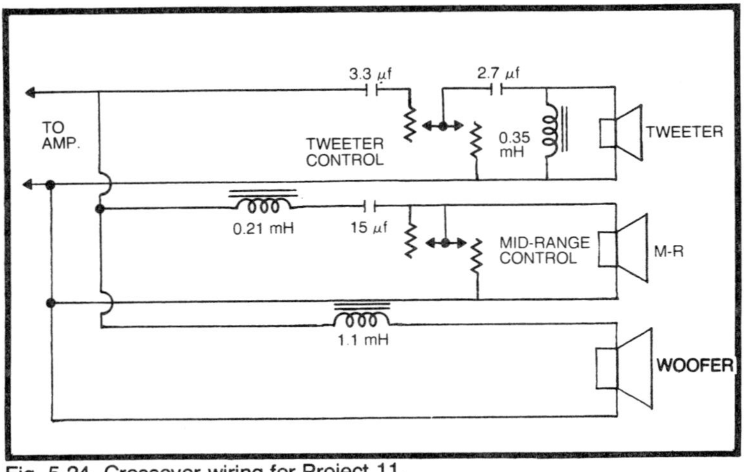

Fig. 5-24. Crossover wiring for Project 11.

Table 5-6. Components parts list for Project 11.

1	12″ woofer, Radio Shack 40-1007
1	5″ speaker, Radio Shack 40-1311
1	Super tweeter, Radio Shack 40-1310
1	Crossover network, Radio Shack 40-1339A Make sure you get the "A" model with crossover points at 1000 and 6000 Hz.

speaker in its hole and turn on your receiver. Turn up the volume just enough to hear the sound easily. Check to see if the lows are coming from the woofer, the mid-range from the middle speaker and the highs from the tweeter. Rotate the controls to see if they control the indicated speakers. When you are satisified that the wiring is right, turn off your receiver and disconnect it. Solder the connections on the speaker terminals. Run a bead of silicone rubber glue around each speaker hole and set the speaker on the glue. Rotate the speaker slightly to spread the glue and press it down hard. Leave the cabinet in this position for at least a half day, preferably a day.

Adding a Grille Cloth

That's it unless you want to add a stretch grille cloth. Make the frame from 1″ × 2″ pine. Paint it flat black and stretch the grille cloth over the front of the frame and staple it on the back. One way to secure a large grille frame is to use #3 or #4-penny finishing nails, driven through the frame at carefully measured points before you install the grille cloth. Then, with the grille cloth on, drive the nails into the front of the cabinet. Use hardwood blocks under your hammer to keep from marking the cloth. Another way to install the grille frame is with several blobs of silicone rubber glue, one at each corner and one at each mid-point of the sides.

If you place this speaker on a spongy carpet, check to make sure the floor under it is level. On a slightly sloping floor there is some danger of the cabinet tipping forward because of the shallow front to rear depth. No problem on a firm floor or a level floor, but you may have to put an extra support under the front foot on problem floors.

When you connect the speakers to your receiver, use care to put the two stereo speakers in phase. Adjust the controls on the back to make the three in-line speakers sound as much like a single source as possible. Then enjoy.

Reflex Speakers

The bass reflex was king of the speakers until the acoustic suspension closed box speaker replaced it. For a generation it seemed that the reflex would just fade away, but in the last few years it has made a spectacular comeback. The gain in popularity for the reflex has occurred because of a new design approach to ported systems, based on what is known as the "Thiele data." What was formerly considered mysterious, almost even magical, has been reduced to a set of data that design engineers can use to predict the performance of a speaker and box combination.

PRINCIPLES OF THE HELMHOLTZ RESONATOR

The benefits of proper reflex loading have been known since the earliest days of high fidelity sound. A ported box corresponds to what is known by scientists as a Helmholtz resonator. Any volume of air in the port vibrates on the cushion of air in the box. When a speaker is added to one wall of the box, things get more complicated. At box resonance the air in the port acts like a piston, vibrating in phase with the speaker cone. As both pistons press against the air in the box together, the piston air acts as a damper on the cone, reducing cone movement and distortion. If the port is smooth and free of any restrictions, the port Q will be high, permitting great air movement through the port. In a well designed reflex the cone is highly damped over a band of bass frequencies at and above box resonance. At these frequencies the port contributes to the bass performance of the system, producing low distortion sound.

THIELE'S DATA AND ITS EFFECT ON REFLEX DESIGN

As compared to a closed box, a reflex system with the same driver installed can give a significantly increased bass range, and at lower distortion. For good reflex performance, a driver should be designed to work in a ported enclosure. Such a driver will have a larger magnet and be more efficient than woofers designed for closed box systems. Although consideration was given to these characteristics of reflex speakers before the data gathered by A.N. Thiele, design work was largely hit or miss. The problems of optimum box characteristics for a given driver, such as box volume or tuning frequency, were debated by different schools of thought. Then Thiele compared the bass reflex to an electrical high pass filter and showed that box volume and frequency should be determined by certain driver parameters, chiefly driver compliance and Q, the degree of magnetic damping on the driver. When a designer follows Thiele's data, he can predict the frequency of bass cutoff. He can also predict the effect of increasing or reducing the size of the box. And, if he is interested, he can even plot a response curve in advance. The curve will show the frequency and amplitude of any bumps or dips caused by compromises made in box design.

In this chapter there are three reflex systems, all based on Thiele data. This doesn't mean that all have optimum volumes behind the cone and perfectly flat response curves down to cutoff. In some cases such performance would require an enormous cubic volume. But each of these enclosures was designed to reach a good compromise between space and performance requirements. There is a subtle difference in the kind of bass response you will get from a ported box compared to a closed box. Several years ago a manufacturer who makes both closed box speakers and ported systems ran a test at an audio show to try to determine whether listeners preferred ported boxes or sealed enclosures. The tests showed that those who heard the comparison were about equally divided in their preference. When asked to comment, the people who like the ported enclosures said they gave a more free and relaxed bass.

LIMITATIONS OF REFLEX SPEAKERS

In spite of its theoretical advantages, the reflex has one trait that doesn't endear it to lovers of music at high sound intensity. This is the phase reversal that occurs below resonance, changing the air in the port from a damper to an uncoupled air mass. This permits the woofer cone to surge at any pulse. Even good quality turntables produce infrasonic pulses that can drive a speaker cone almost out of its frame. Or tuning an FM tuner from station to station can produce the same symptom. The cure is to use the low-cut filter on the amplifier or receiver. Such filters remove no usable bass, only the sub-bass sounds you don't hear, but they can overload a vented woofer. In fact, when you switch in a properly designed infrasonic filter, you can hear the bass grow firm. Contrary to expectations, the speaker will appear to have a deeper bass. The ultimate solution to the cone surging problem, which is worse with some kinds of drivers than others, is given in the next chapter. It is the bass reflex equalizer. But none of the three speakers described in this chapter produced an above normal susceptibility to surging.

If you like superior bass response, you should try a reflex speaker at least once. Here are three for your inspection.

PROJECT 12: THE ENERGY SAVER

- **Construction:** Use power saw.
- **Wiring:** Homemade crossover network.
- **Comments:** Good efficiency and dynamic range. Musical sound.

In one listening session with this speaker someone made the comment that the bass was a bit too prominent. At that moment the speaker was sitting on the floor near a corner of the room. Someone then picked the cabinet up and placed it on another speaker, raising it about 18″ above the floor. The same listener who had commented on the prominent bass came back and, hearing the speaker in its new position, said, "That is one of the most natural and musical sounds I've ever heard."

This incident shows the importance of room placement. Even the most perfect speaker can be ruined by improper location. Obviously it pays to start out with the best performing speaker you can afford, but careless attention to such details as room acoustics can make the true quality of your speaker irrelevant.

ADVANTAGES AND DISADVANTAGES

This project is a particularly good choice if you have an amplifier of limited power output (Fig. 6-1). Compared to many speakers it has the ability to double the effective power of your system. But if you do have an amplifier with lower power capability, don't make the mistake of driving any good speaker too hard with it. Low power amplifiers have a tendency to clip, changing sine waves to square waves. This trait can destroy tweeters even though the tweeter is rated at a higher power than the amplifier.

Fig. 6-1. Drivers with enclosure of Project 12.

Even if you have an amplifier of adequate power output, this speaker system offers another rare advantage that you probably haven't considered—a chance to reduce your electric bills. Let's face it. Energy hungry speakers do require more power, which demands more current from your utility company. The difference in kilowatts may not be startling. After all, though, if it's worthwhile to choose smaller light bulbs, shouldn't you also conserve on amplifier power? This system consists of a full range speaker that is at least twice as efficient as most current 8″ speakers and a small dome tweeter in a ported box. Although the full range speaker boasts a wide frequency range, the addition of the small tweeter gives improved high end response and dispersion.

Before you rush out to buy the components for this system, there are some points you should consider. First, reflex systems are slightly more contrary than closed box speakers. This means that you must get the enclosure volume and tuning right. Sloppy work can ruin any speaker system, but unplanned leaks can detune a reflex with totally unpredictable results. Also, as mentioned earlier, check to see if your receiver or amplifier has an infrasonic or "subsonic"-filter. In a normal size living room, even without the low frequency filter, you may also have no trouble unless you like to blast down the walls with your audio system. But since many listeners seem determined to do just that, you should be forewarned. Of course, if your receiver has the power to damage a woofer, it probably has the filter also.

Most speakers are designed with a gasket at the front so they can be installed behind the speaker board, but the flat rim behind the gasket permits easy front mounting. The *Electro-Voice MC8A*, used in this project, is an exception. Because of its unusual frame design, this speaker must be installed behind its mounting board. When used with a front-mounted tweeter, the normal difference in driver depth, and therefore voice coil placement, is exaggerated by the thickness of the speaker board. Here the distance between voice coils is reduced by using a two-piece board, made up of a sheet of ¼" plywood glued to a sheet of ¾" particle board. The speaker is installed on the ¼" board, but that board is reinforced by the thicker board over its entire surface except on the small area occupied by the speaker gasket. This multiple panel speaker board is the only unusual enclosure feature in the project.

TWEETERS AND TWEETER CONTROL

Dome tweeters can be damaged if subjected to drive voltage at frequencies below their optimum range. The crossover network used in the project was designed to protect the tweeter by sharply cutting off energy to it below about 3000 Hz. By using a 3 element network in the tweeter leg of the circuit, the roll-off rate of the tweeter is 18 dB/octave. To simplify crossover construction, the woofer's cutoff rate is much slower at 6 dB/octave. A resistor and capacitor across the woofer equalizes its impedance over and beyond its operating range.

In the first stages of project planning it was expected that no tweeter control would be necessary because of the relatively high efficiency of the woofer. When the first speaker was put in use, listeners commented on the lively highs, and then suggested that they should be tempered a bit. After an L-pad was installed, most listeners agreed that is was a necessary, or at least highly desirable, addition.

If you have never wound a coil, you may hesitate to make your own crossover network because of lack of experience or testing

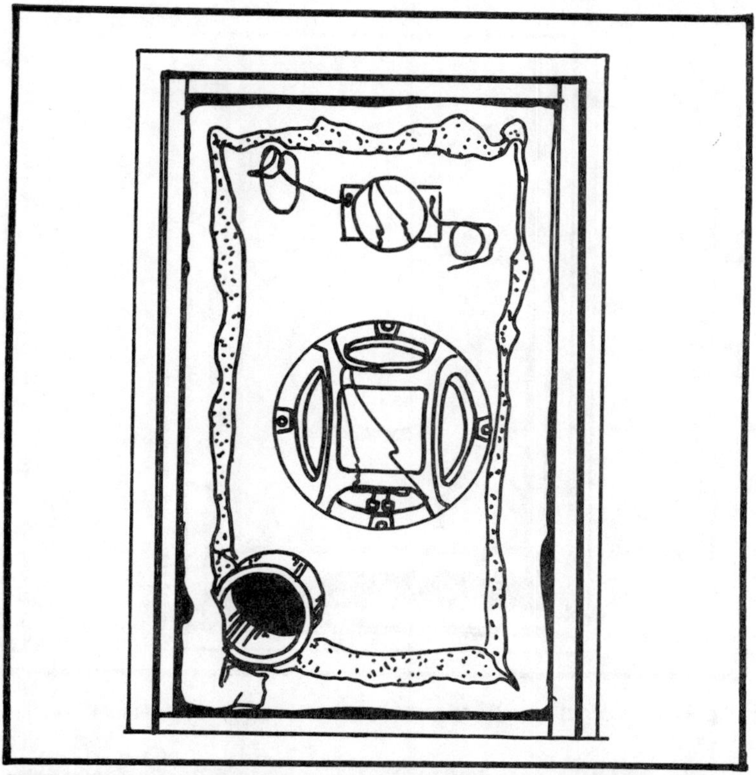

Fit. 6-2. Enclosure with drivers,vent tube and fiber glass insulation installed. Vent tube was later shortened to the tune box to the driver.

equipment. Don't let either reason stop you. If you make the coil forms to the indicated dimensions and carefully measure the wire length, you can wind the coils with confidence that the final value will be very close to that specified. The only precaution that you must take is to maintain enough tension on the wire to produce tight, close fitting turns. It isn't necessary to count the turns. While most of the hand-wound coils shown in this book were made on temporary forms that could be taken apart after winding, the coils shown here were wound on permanent wood forms that were installed in the crossover network. Either method is satisfactory.

CONSTRUCTION

Because of the requirement for rear-mounting of the 8″ speaker, the enclosure should be made with a removable back (Fig. 6-2). This was accomplished by using ¾″ plywood of very high

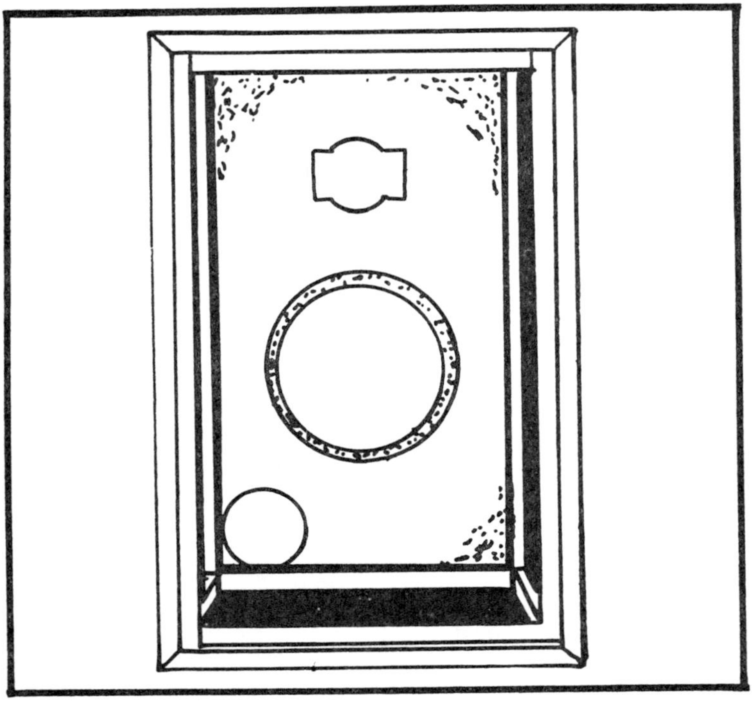

Fig. 6-3. Internal construction of enclosure. Note countersunk woofer hole.

quality with beveled joints at the box corners (Fig. 6-3). Note the specification of high quality plywood (Fig. 6-4 and Table 6-1). If you can't get top quality plywood, or feel you can't afford it, you should enlarge the box 1″ in each dimension and line the interior with ½″ particle board. Or you can build the box by using ¾″ particle board put together with butt joints. Regardless of the method of construction, make sure the internal dimensions are identical to those shown in Fig. 6-3. Remember that any variation in cubic volume can change the tuning.

Cutting Parts

When you cut out the parts, you must cut holes in the two panels that make up the speaker board that are of different diameter, for the 8″ speaker, and even different shapes, for the tweeter. The port opening must be smooth and of equal diameter right through the board, so don't cut it until the two parts are glued together. It's a good idea to postpone cutting the speaker board

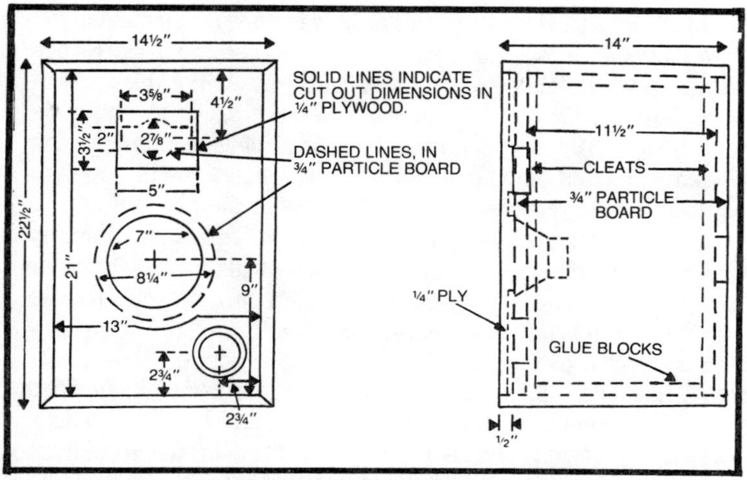

Fig. 6-4. Construction plans for enclosure of Project 12.

panels and the back until you have assembled the box shell so you can make those parts fit perfectly.

Cut the plywood walls to width, then make the bevel cuts for the joints. Set your saw at 45 degrees and test it on some waste material. Set the cut pieces on edge on your work bench and fit them together to see if the joint is close and if the pieces make a square corner. Make any necessary minor adjustments and cut the four pieces to their correct length.

Table 6-1. Materials list for enclosure of Project 12.

¾" Birch plywood, good quality, pieces needed:		
2	14-½" × 14"	Top & bottom
2	14-½" × 22-½"	Sides
¾" Particle board, pieces needed:		
2	13" × 21"	Speaker board and back
¼" Plywood, one piece needed:		
1	13" × 21"	Faceplate
⅜" Plywood, one piece needed:		
1	12-15/16" × 20-15/16"	Grille frame, cut out to leave a frame 1" wide on each side.
Tube for vent, one needed:		
1	⅞" length of 3" I.D. tube. Or substitute a vent of square cross-section with internal dimensions of 2-21/32" × 2-21/32". and same length.	
¾" wood veneer stripping.		
1	78" length to cover plywood edges.	

Mark lines inside each panel to show the proper position of the cleats that act as stops for the speaker board and the back. Notice that the cleats should be indented ¾" at the back, the thickness of the back panel, but 1-¼" at the front to leave space for the grille frame (Fig. 6-5). Apply a sparse coating of glue to the matching beveled joints and put the shell together. Use straps or ropes to hold the parts together under pressure while the glue sets.

Cut the cleats to length from ¾" × ¾" material. Start by cutting the side cleats and installing them with glue. Use clamps or #8 × 1-¼" wood screws through the cleats into the walls to hold them while the glue sets. Measure the length of the remaining cleats carefully so they will fit in the space between the cleats already installed. Minor gaps of, say, ⅛" or less can easily be filled with caulking compound, but try to get the fit exact. After installing these cleats, measure the distance in the corners between the front and rear cleats and cut glue blocks from ¾" × ¾" material to fit into these spaces. Apply glue to the blocks and the interior corner surfaces. Then clamp or screw the blocks into place. Glue blocks will add considerable strength to the beveled joints.

Cut the speaker board panels and the back panel to fit closely. If your box is slightly out of square, you can compensate for that, and then mark the top or bottom of each piece to indicate which is the outer surface. Cut the speaker holes in the inner speaker board panel, the thicker member. You can easily get the tweeter cutout right by first making a circular hole with a diameter of 2-⅞" diameter, centered in the cutout, and then the horizontal rectangular cutout. After you have the speaker holes cut in the inner board, check to see if the 8" speaker frame will just barely pass through the hole. Then clamp the two panels together and mark the outlines of the speaker holes on the outer piece of ¼" plywood. Separate the panels and draw several diameters through the marked 8-¼" circle on the plywood to draw a 7" circle concentric to the larger one. Cut out the 7" circle. Draw vertical and horizontal lines through the marked outline of the tweeter hole. Using these lines as a guide, mark the outline of the tweeter cutout in the plywood panel. After you make this cutout, use a wood rasp or sandpaper to round off the outer edge of the tweeter opening in the plywood panel.

Coat the matching surfaces of the two speaker board panels with glue and press them together. Apply pressure by laying the boards on a flat surface and weighting them. It's a good idea to put several flat panels on top of the speaker board, then the weights. Or you can clamp the parts together to apply pressure. When the glue

Fig. 6-5. Grille frame, painted black here, was cut from 3/8″ plywood.

has set, place the 8″ speaker in its proper position and mark the location of the mounting holes. Drill the four 3/16″ holes and install four round head 3/16″ × 1″ bolts in them. Paint the front of the board flat black. Then glue and mount it in the enclosure, using furniture clamps or screws to hold it until the glue sets.

Applying Caulk

Cover the raw front edge of the plywood walls with thin wood veneer that matches the wood of the sides. Sand the cabinet exterior and apply the shade of stain that you desire. Then apply a dull varnish or an oiled finish to the wood.

Caulk the interior joints with a latex or silicone rubber caulk. Place the box face down and install the 8″ speaker. Use only moderate pressure on your wrench to tighten the nuts; the gasket will form a proper seal to the board without excessive force that could deform the speaker frame. This precaution should be followed when bolting in any speaker; the MC8A has a die-cast frame that is less likely to be twisted than most others.

It's a good idea to solder "pig-tails" to the tweeter before installing it so the final connections will be easier to make. Cut the leads long enough to reach the lower back of the enclosure. Note carefully which one goes to the positive pole of the tweeter, the one with the red insulator under the terminal. Set the cabinet on its back and run a bead of silicone rubber caulking compound around the tweeter cutout in the particle board panel. Set the tweeter on the board and press it down to make the rubber form an air-tight gasket. Leave the box in that position while the caulking material is setting, preferably a half day or more.

PREPARING THE CROSSOVER NETWORK

Next, prepare the crossover network. Make up two coil forms, using a ¾″ diameter wood dowel and square pieces of masonite or thin plywood. Cut the dowel in ⅜″ lengths. Glue and clamp pieces of thin plywood onto each dowel to form the sides of the form. If you have no clamps, you can use brass screws to assemble the forms. Don't use iron screws or the inductance will be increased beyond the predicted value.

When the forms are ready to use, drill a small hole in one side of each form, located adjacent to the dowel. Measure out the correct length of wire and feed one end through the hole in the side of the form with about 3″ of wire outside the form. Start winding the coil, keeping each turn tight. When you have the last turn in place, hold the wire taut until you, or a helper, can wrap the coil wire with tape to hold the strands in place. If you follow directions and get the length right, your coils will have a value of inductance very close to that specified. If you have access to an impedance bridge, or other test equipment, you can add several feet to the indicated length; then remove enough turns to get the value exactly right. But you will probably not be able to notice the difference that such precise coil trimming will make in the sound of your speaker system.

Cut an 8″ × 8″ crossover board and drill it for the L-pad and the speaker screw terminals (Fig. 6-6). Install the L-pad with its hardware. Mount the speaker terminal board with glue and tacks or small screws. Mount the coils and the other terminal strips with Elmer's Heavy Grip Cement or Leech's Heavy Adhesive. Then wire the crossover and solder all connections. In wiring it, pay special attention to getting all the insulation removed from the ends of the coil wire. Most insulation on wire in current production can be

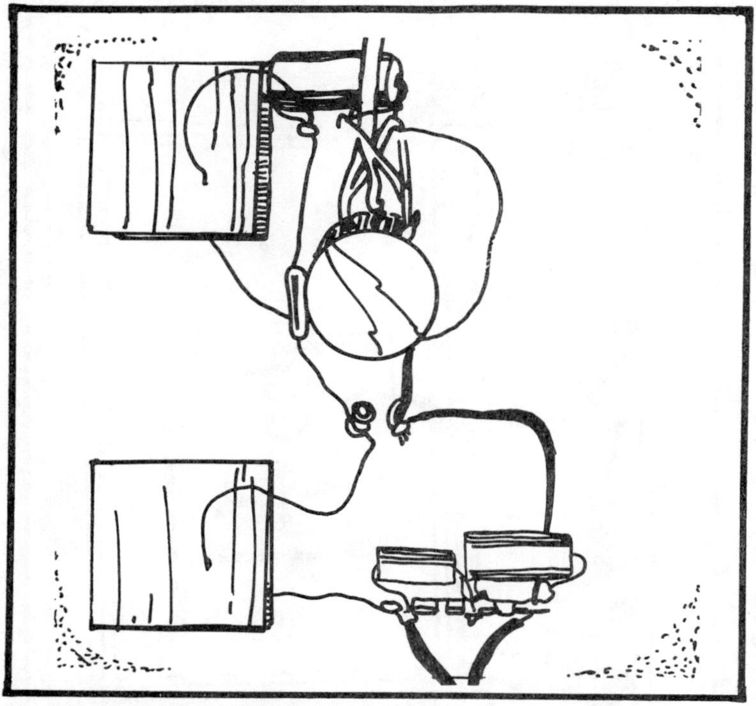

Fig. 6-6. Parts layout on 8″ × 8″ crossover board.

removed with a soldering gun or iron. Or you can scrape or grind the tough material away to expose the raw copper. Failure to get a good connection for the coils will defeat the proper operation of the network.

Cut a 4″ × 4″ hole in the enclosure back, centered behind the position of the woofer in the front of the box. Run a bead of silicone rubber around this cutout, placing the sealer about 1″ from the edges of the hole. Then press the crossover board down on the rubber compound. Leave the back in a horizontal position until the rubber sets.

Line the interior walls of the enclosure with damping material. In the system pictured here, a 1″ thickness of acoustical fiber glass was used on every inside surface except the front. Then a 3″ thick blanket of damping material was added to the interior of the back panel (Fig. 6-7). Keep the damping material out of the port. Cut a ⅞″ length of tubing and glue it in the port hole to tune the box.

Install strips of ¼″ × ⅜″ open cell foam weather stripping material to the inside of the back panel, just inside the screw line.

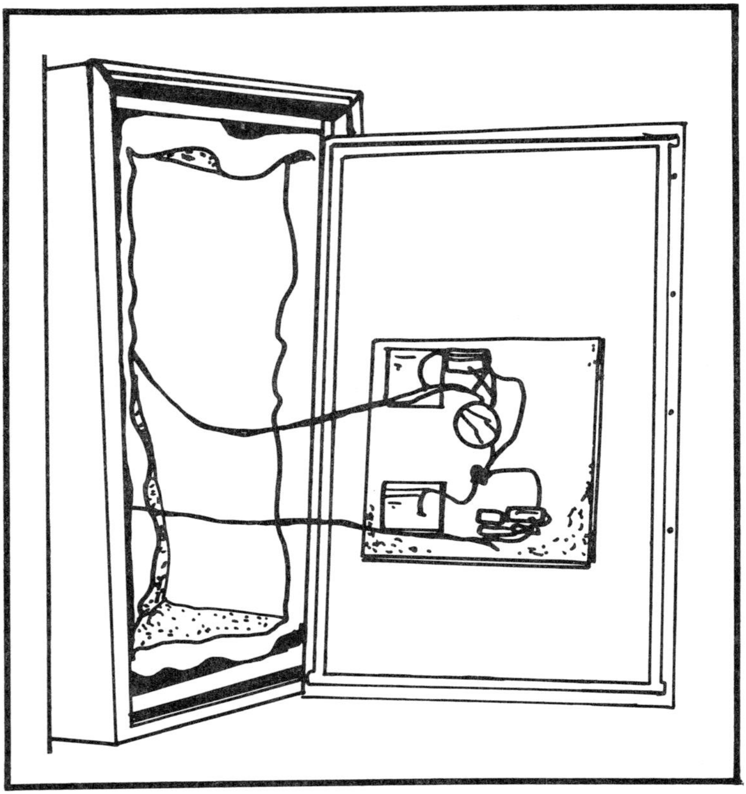

Fig. 6-7. Use foam gasket around the back, but inside the screw line, to seal the enclosure. Don't forget to cover the back with damping material.

Connect the leads from the tweeter to the crossover network and the woofer leads from the crossover network to the screw terminals on the woofer (Fig. 6-8 and Table 6-2). Notice that the tweeter is wired out of phase with the woofer. This is done because of the different degrees of phase shift in the crossover network for the woofer and the tweeter. For the woofer leg of the circuit the phase shift is 90 degrees, for the tweeter, 270 degrees. Thus the net difference is 180 degrees. This difference of phase shift, 180 degrees, means that the sound from the woofer is out of phase with that from the tweeter at the crossover frequency. If the two drivers are wired with the same polarity, there will be cancellation of sound in the band where sound from the two drivers overlaps and where output of the two is similar in loudness. This produces a hole, or at least a dip, in the response curve of the system.

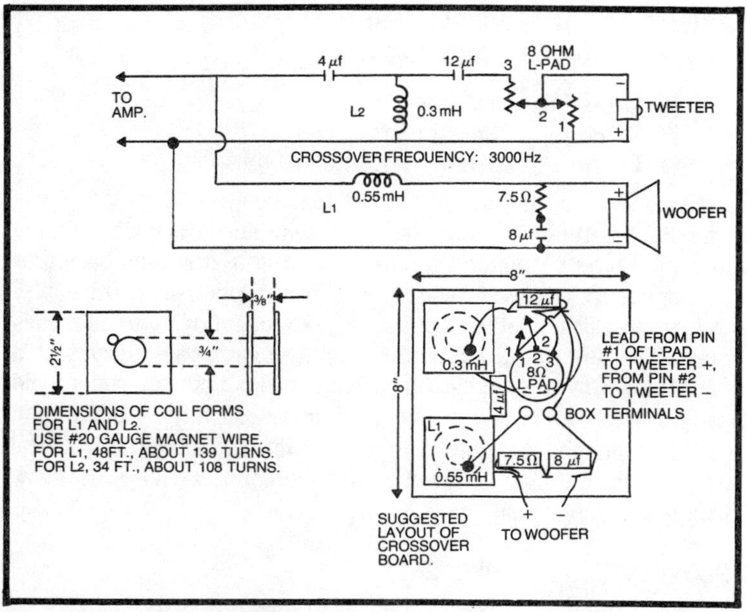

Fig 6-8. Crossover wiring for Project 12.

The last step is to screw down the back and connect your speaker to a receiver or amplifier. Don't be surprised if this system appears to add muscle to your small receivers or amplifiers of modest power. And, in these days of the energy crisis, it's good to know that you can save a few pennies on your electric bill. Plus, your receiver can coast along more of the time, drawing less current from your utility power line. There are probably a half dozen, maybe more, speakers in this book that cost less to build than the 2-way reflex. But when you consider the long term inflationary trend of the price of electric energy, this just could be the bargain of the group.

Table 6-2. List of components needed for Project 12.

1	8″ full range speaker, Electro-Voice MC8A
1	Dome tweeter, Peerless K01DT
1	8 ohm L-pad
1	4 μf non-polarized electrolytic capacitor
1	8 μf non-polarized electrolytic capacitor
1	12 μf non-polarized electrolytic capacitor
1	7.5 ohm, 10 watt resistor
1 lb.	#20 gauge magnet wire. (For 2 Speaker systems)

PROJECT 13: THE DOPPELGANGER —A MASS-LOADED BASS RADIATOR

- **Construction:** Easy.
- **Wiring:** Homemade crossover.
- **Comments:** Deep, fat bass. 4 ohms.

This system has several novel features, all adopted for the purpose of getting the best sound possible at a low cost. It has a full-bodied bass response without the one-note imitation bass that some speakers produce. While the full bass response is noticeable, the system will accept bass boost without an annoying middle bass boom. This is evidence that the slightly plump bass occurs at a comfortably low frequency. As with any reflex speaker you should use an amplifier or receiver that has an infrasonic filter. This caution is worth repeating, plus the fact that such filters don't cut off musical bass. When the filter is switched in, you can hear the bass grow cleaner and more solid.

UNCONVENTIONAL COMPONENTS

While the outstanding feature of this system is its full bass, it also boasts a response into the ultrasonic range via a super tweeter that fires its highs at the ceiling. There are two reasons for the unusual tweeter angle. First, it saved a tweeter control that would have been necessary if the tweeter had been mounted on the front of the enclosure. And listeners judged the highs to be more natural from the tweeter in that position. This unusual tweeter placement also affected one feature that meets the eye. With a grille cloth-covered front and top, plus slab sides, the styling is a change from the typical speaker box (Fig. 6-9). If you greet its appearance as a refreshing alternative to the usual boxes, give it a try.

One of the unusual features of this system that you don't see is the highly unorthodox crossover network. It is based on a network described in a scientific journal several years ago by Ashley and Kaminsky. They called their network a "quasi second order" network because, while it required only as many components as a first order 6 dB/octave network, it offers a cutoff rate of 12 dB/octave at the crossover frequency with the rate shifting to 6 dB/octave beyond that point.

The "quasi second order" description applies to the circuit that divides the spectrum between the woofer and mid-range driver. These speakers each have a 7-ohm impedance at 1500 Hz, the desired crossover frequency. A typical crossover network would

Fig. 6-9. Project 13, with slab sides and upward firing tweeter, gives an unconventional appearance.

employ components with reactances of 7 ohms at that frequency, but here the reactance is reduced to 3.5 ohms at the crossover frequency. Because of this requirement for reduced reactance, the value of the choke is halved and the value of the capacitor is doubled as compared to the values of components in a typical 6 dB/octave filter.

Note that the choice of 1500 Hz was not arbitrary. At that frequency the impedances of the woofer and mid-range speaker were equal, which is desirable for a series connection such as this one. Also, 1500 Hz was considered a desirable choice for the low-cost mid-range driver. This driver has a resonance of about 700 Hz. With the quasi second order network, the 1500 Hz frequency crossover effectively reduces output at resonance and increases the power handling ability of that driver. While the super tweeter needs no crossover network to protect it, a capacitor-resistor filter cuts the power to it below 4000 Hz at the rate of 3 dB/octave.

The extra bass cone, sometimes called a passive radiator (PR) or auxiliary bass radiator, is mass-loaded to tune the system to

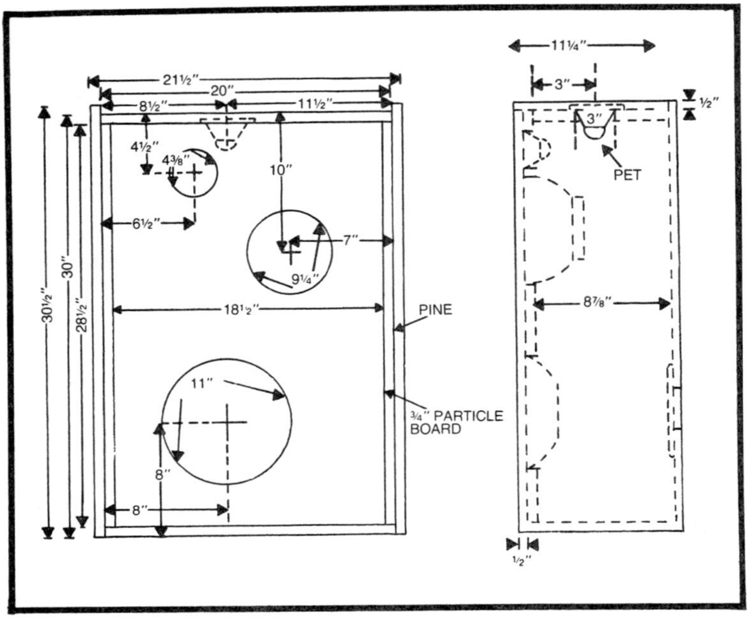

Fig 6-10. Construction plans for enclosure of Project 13.

slightly above 30 Hz. This tuned frequency was predicted by tests made on the bare 10″ woofer and confirmed by the shape of impedance curves run on the completed system. So the list of unconventional components includes the crossover network, the large bass drone cone, the piezo-electric super tweeter, and the upward firing placement of that tweeter with its effect on the styling. All in all a very unusual speaker system, but one which listeners have applauded, possibly because it is different.

BUILDING PROCEDURE

The construction is simplified by making only two parts visible in the completed enclosure—the exterior sides. Except on those two strictly ornamental panels, the grille cloth will hide all but the crudest workmanship. The only requirements for a professional looking job is that you use normal care on those sides and that you flush mount the speakers so the grille cloth will lie flat. If you have a router, you can rout the outer edge of each speaker hole deep enough to permit the speaker to mount with the edge of the gasket flush with the front of the baffle. If you have no router, you can easily accomplish the same thing by gluing an extra piece of ⅜″ plywood,

Table 6-3. Materials list for enclosure of Project 13.

¾″ Particle board, pieces needed:		
2	8-⅞″ × 20″	Top & bottom
2	8-⅞″ × 28-½″	Sides
2	20″ × 30″	Speaker board & back
⅜″ plywood, one piece needed:		
1	20″ × 30″	Faceplate
1″ × 12″ pine boards, pieces needed: (actually ¾″ × 11-¼″).		
2	11-¼″ × 30-½″	Exterior sides.

cut out to accommodate the outer diameter of the speakers, to the front of the ¾″ speaker board. This is the method shown in Fig. 6-10. Table 6-3 lists materials. To flush mount the super tweeter in the top panel, you can use a wood chisel or other sharp tool (Fig. 6-11).

Cut out the speaker holes at the prescribed locations before assembling the box. A stereo pair should be made as mirror images of each other, so reverse one speaker board when you put the cabinets together. Locate a 3″ × 4″ cutout for the crossover board

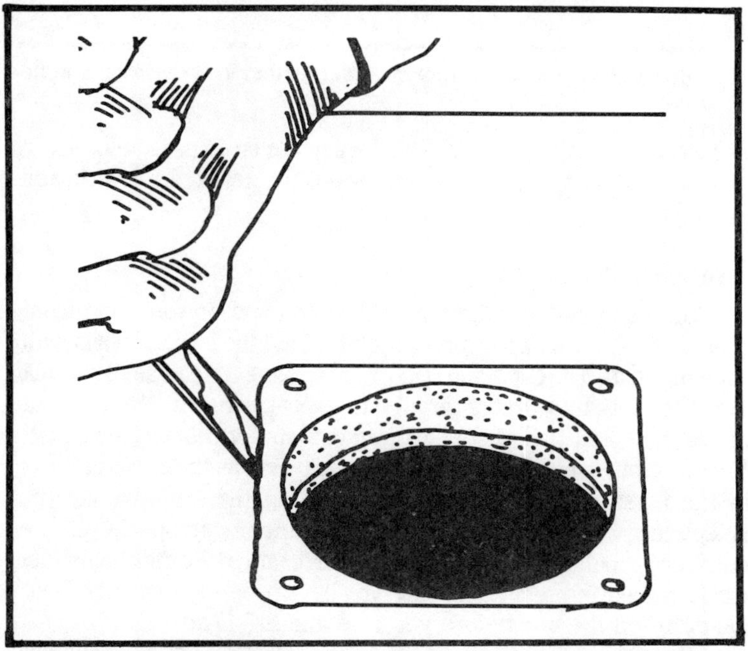

Fig. 6-11. To prepare top board for super tweeter, cut a 3″ hole, set the tweeter in the hole and mark the outline on the board. Remove about 1/16″ of material to permit the tweeter to fit flush.

6-12. After assembly of the main box, glue and nail the pine exterior sides to the enclosure.

in the back panel. If you place this behind the large passive radiator cone, you will have easy future access to the network for wiring or trouble shooting.

Assembling the Box

Glue and nail together the sides, top and bottom, and front speaker board. Caulk the interior joints. Line the interior walls with damping material. Glue and nail the rear panel to the sides. Caulk the rear joints by working through the speaker holes. Next cut out the ⅜″ faceplate and make enlarged cutouts for the speakers in it. These cutouts are centered on the same points as the speaker holes in the inner speaker board. After making the cutouts, lay the faceplate in place on the box and check the fit of your speakers. The speakers should lie flat on the inner board, with the outer board just clearing their frames. Make sure you have properly located the ⅜″ faceplate on the box before you glue and nail it down.

Paint the front and top of the box flat black. When the paint has dried, install the external slab sides with nails and glue (Fig. 6-12). Coat the entire surface of each side of the box with glue before nailing on the external side. Just a few nails around the edge will hold the panel until the glue sets up. Stain and finish these sides; then fill the nail holes with a color-matching putty before installing the speakers.

Mount the crossover components on an 8″ × 8″ piece of ¼″ tempered hardboard, according to Fig. 6-15A. Measure out enough lamp cord to reach from the crossover network in the lower rear of the box to each of the three drivers, about 36″ for the mid-range and tweeter and 24″ for the woofer. Connect the leads to the crossover and solder the connections. Then glue the crossover board over the hole in the back. Cover the back panel with 2″ to 3″ of damping material, bringing the speaker leads out through holes in the material.

Wiring the Mid-Range Speaker, Woofer and Tweeter

Put the enclosure on its back and wire the mid-range speaker and the woofer. Install each by running a bead of silicone rubber glue

Fig. 6-13. System and drivers. There is a double thickness of 1/2″ washers glued to the passive radiator cone, making eight in all.

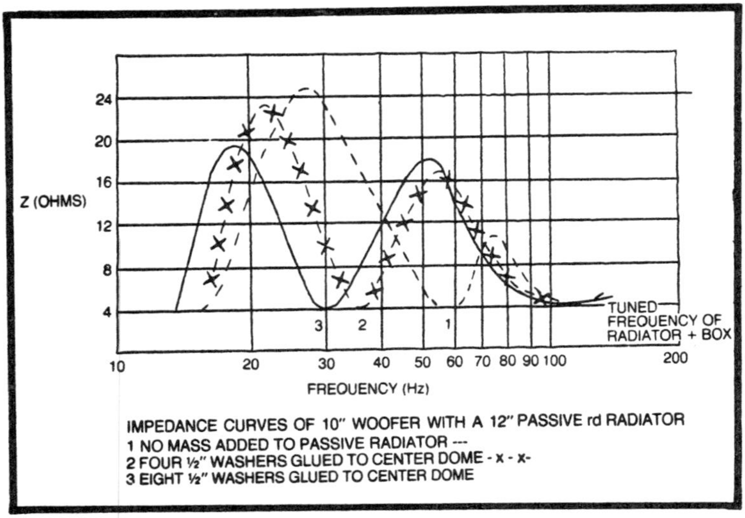

Fig. 6-14. Here's how loading the passive radiator cone alters the impedance curve of the woofer in Project 13.

around the mounting holes and pressing the speakers down on the glue. Install the bass radiator the same way. Use a small amount of silicone rubber to glue eight ½" washers on the dust cover of the bass radiator. These should be put on in a symmetrical pattern, two deep. Check Fig. 6-13 for a suitable pattern. Also see Fig. 6-14.

When the glue has set, place the cabinet in an upright position to wire and install the tweeter. Silicone rubber doesn't adhere well to the smooth plastic tweeter, so it's a good idea to use four #8 × ½" sheet metal screws for a more secure mounting.

Check the speakers for proper wiring before installing the grille cloth (Fig. 6-15 and Table 6-4). Listen to each driver to see if the lows come from the woofer, the middle frequencies from the mid-range driver, and the upper highs from the tweeter. Adjust the mid-range control to test it; then try to balance the sound from the woofer and mid-range driver.

Attaching the Grille Cloth

Cut a piece of grille cloth just wide enough to cover the front of the cabinet, excluding the slab sides, when stretched. This piece should measure a little less than 20" wide and about 42" to 45" long. Place the grille cloth on the box so the cloth overlaps the bottom edge by an inch or two. Staple this end of the grille cloth first. Set the cabinet upright and grasp the middle of the top edge of the cloth.

Pull it upward until it is taut; then bend it over the top and down the back. Drive in a few staples near the middle of this edge. Stretch the cloth horizontally and staple the edges of the front and top. Finish stapling the back corners. Check the grille cloth for straight threads and make final adjustments. Staple one end of a piece of matching ribbon, about ¼″ to ⅜″ wide, on the bottom at one edge. Run the ribbon up the edge of the grille cloth so it covers the cut edge; then stretch and staple it at the back. Repeat this for the other edge.

Position the speakers in your listening room so that the off-centered mid-range drivers are on the inside of the triangle made by the enclosures and the listener, the side nearest the other speaker. If you are using the speakers in a small room, where the distance between them is, say, 5′ or less, try switching positions so the mid-range drivers are separated by the greatest possible distance. When you find the proper location for your speakers, you will find that they offer a quality of sound that is amazing, especially considering their low cost.

PROJECT 14: THE TITAN

- **Construction:** Fairly easy, except for joining boards.
- **Wiring:** Homemade crossover.
- **Comments:** Effortless bass.

Here is a speaker system for the big bass freak (Fig. 6-16). Not everyone will leap at the prospect of putting two 150 pound cabinets in the living room, even if the cargo inside includes the kind of bass that only a big box can give. But if you have the space and a firm foundation, this could be your boxcar.

As you may guess, in planning this project the emphasis was put on bass response. And that is the distinguishing feature you

Table 6-4. Speakers and components needed for Project 13.

1	12″ passive radiator, McGee Model 12PR
1	10″ woofer, Cletron model CIC 100
1	5″ mid-range speaker, Panasonic EAS-12PM33SB
1	Super tweeter, Motorola 59067 superhorn, piezoelectric. Speakers available from McGee Radio Co.
1	8 ohm L-pad.
1	30 μf non-polarized electrolytic capacitor (or any combination of capacitors to make 30 μf.)
1	5 μf n. p. elect. capacitor
1	8 ohm resistor, 2 watts or greater.
¼″ lb.	#20 gauge magnet. wire. Note: ¼ lb. will make coils for two speakers.

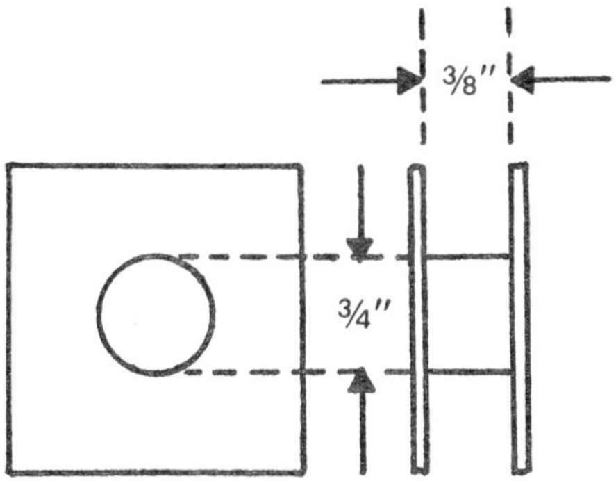

DIMENSIONS FOR COIL FORM.
USE #20 GAUGE MAGNET WIRE.
LENGTH, 38½′
ABOUT 117 TURNS.

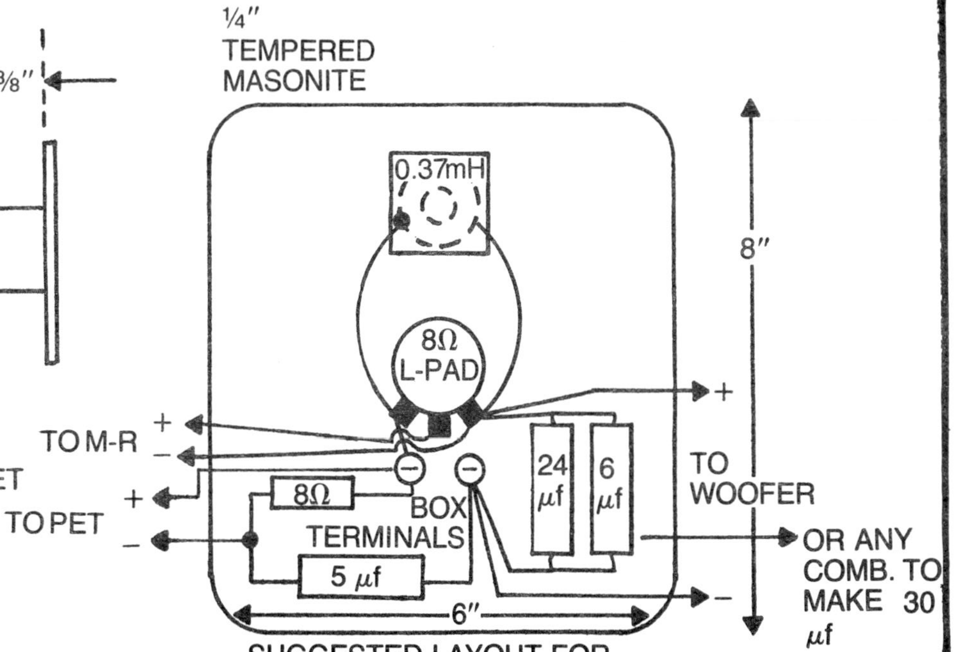

SUGGESTED LAYOUT FOR CROSSOVER BOARD.

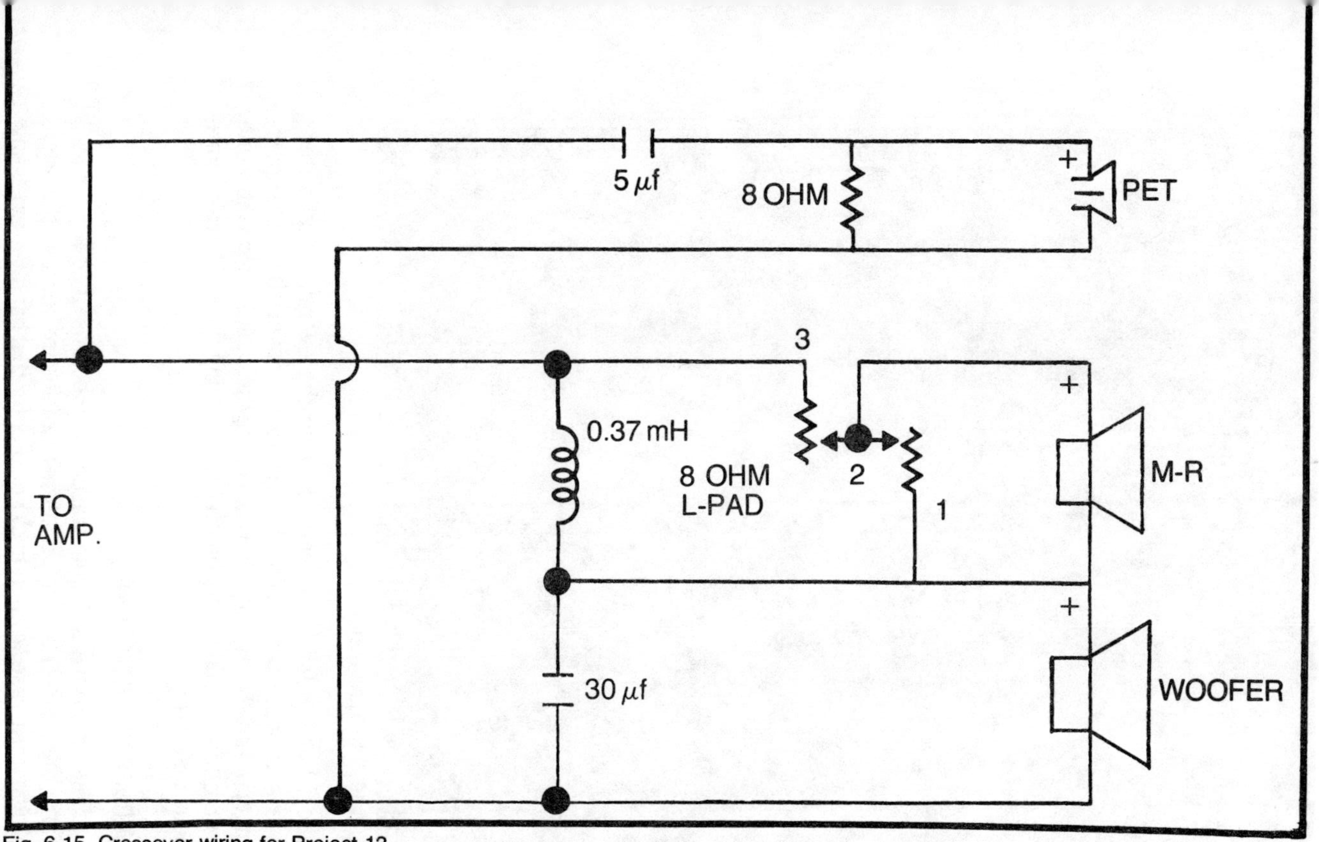

Fig. 6-15. Crossover wiring for Project 13.

Fig. 6-16. The Titan at home.

notice, but the full bass seems to give the overall sound a mellowess. In some locations the bass can almost overwhelm the upper range; in such cases you can either move the cabinets for a different room effect or add some length to the tuning duct. There's more about that later.

MAKING THE ENCLOSURE

The enclosure is double-walled, pine boards over ¾" particle board. Figure 6-17 shows a mid-range enclosure because the original plan was to use an open-back speaker for that band. After testing several mid-range speakers, a closed-back special mid-range driver was chosen, so you can forget the mid-range box. Once you install this speaker, you won't want to be moving it out to get to rear-mounted controls. The mid-range and tweeter controls were installed on the speaker board (Fig. 6-17). See Fig. 6-18 and Table 6-5.

Cut out the particle board panels first. Make the cutouts in the speaker board and set it aside. Drill two ¼″ holes in the center of the back, spaced to match the lugs on a screw terminal strip. You can mount the strip later. Mount at least two 2″ × 2″ or 2″ × 4″ braces

Fig. 6-17. Layout of parts on back of the speaker board. Mid-range box, shown at right was planned for the open back mid-range driver which was not used.

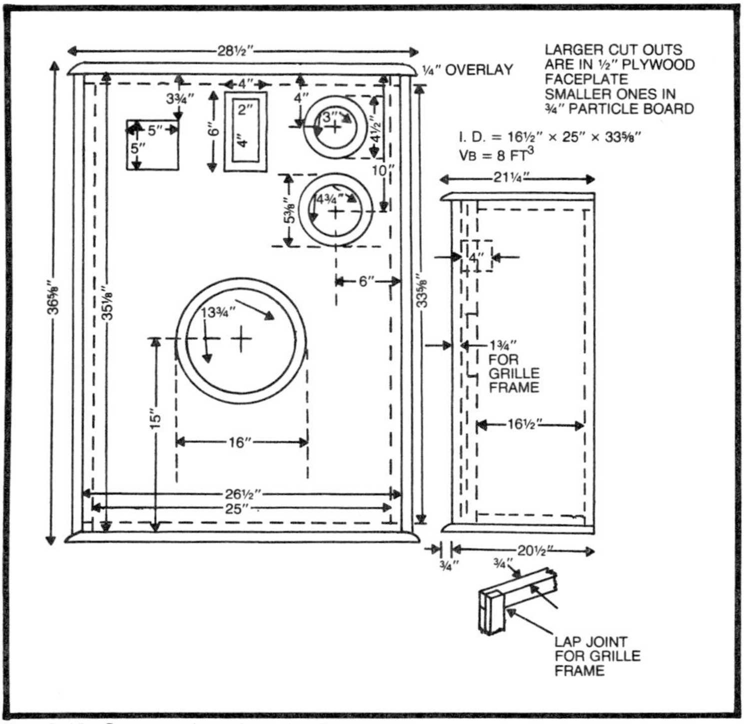

Fig. 6-18. Construction plans for Project 14.

on the interior of the back. These braces should be about 32"to 34" long, installed vertically with glue and nails.

Main Box Assembly

Assemble the main box by nailing the bottom to the sides, then the top. To help hold the sides in position you can lightly nail the back onto the sides, but use no glue there (Fig. 6-19). After putting together the shell, you can set the box on its face to glue and nail down the back. Cut and install 2" × 2" cleats for screws at the front of the box, making these set flush with the front edge of the particle board sides (Fig. 6-20).

Prepare the pine exterior by cutting pieces of 1" × 12" pine to the correct length and width. Note that you must join two lengths of pine to make the 21-¼" wide pieces for the top and bottom and the 20-½" pieces for the sides. Pine boards sold as 1" × 12" are really only about 11-¼" wide, so very little must be ripped from the full

Table 6-5. List of enclosure materials for Project 14.

¾″ Particle board, pieces needed:		
2	16-½″ × 26-½″	Top & bottom
2	16-½″ × 33-⅝″	Sides
2	26-½″ × 35-⅛″	Speaker board & back.
¾″ Pine, pieces needed: (made up by joining 1 × 12′s)		
2	21-¼″ × 28-½″	Top & bottom
2	20-½″ × 35-⅛″	Sides
½″ plywood, needed:		
1	26-½″ × 35-⅛″	Faceplate.
2	6″ × 2-¾″	To make duct
2	5″ × 2-¾″	behind vent.
1″ × 2″ Pine:		
1	26-7/16″	Grille frame; use lap joints
1	35-1/16″	if possible.
Molding or picture framing material, as desired:		
1	26-7/16″	Facing for grille frame.
1	36-1/16″	

Fig. 6-19. Begin construction by assembling particle board inner box. Here the back is used to temporarily hold sides while the top is glued and nailed on.

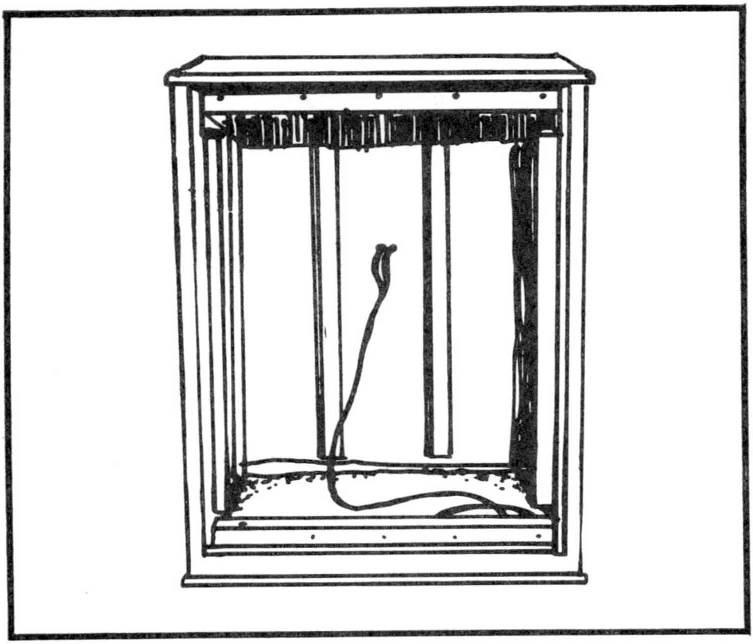

Fig. 6-20. Internal construction of box. The 2″ × 2″ cleats are for screws only and don't have to fit tightly. The box is sealed by a foam gasket that fits against the front edge of particle board box walls.

width for the top and bottom pieces. If you have access to a joiner, you should run the boards through the joiner first to get a close fitting joint, and then rip the rear board to the proper width.

Glue the pine boards together to make the wide panels. The top of the enclosure makes a suitable work surface for this job because it isn't wide enough to interfere with the projections on bar clamps (Fig. 6-21). Preferably you should do this gluing on the same day you install the pine on the box, so that the pine boards won't warp. The procedure is to glue all the boards. After waiting a few hours for the glue to set, rout the sides and front edge of the top and bottom to give a smooth, rounded appearance. Then install the sides, bottom, and top on the box with glue and nails.

If you don't have enough clamps to clamp all the boards at once, start with a side. Then, after you mount that side, do the other side. Next do the bottom and then the top. Regardless of whether you make up the panels at one time or in sequence, you should follow that same order in mounting the parts on the particle board box. And don't forget to rout the bottom before attaching it to the box. If you don't have a router, you can use a rasp and sandpaper to give a similar effect.

If you have no bar clamps, you can probably get satisfactory joints anyway. Glue and nail down one board. Then glue and nail down the adjacent board, holding it tightly against the first one. This method will produce a coarse joint, but your cabinet can still be attractive and just as solid.

Finishing the Cabinet Exterior

You can now finish the cabinet exterior, but first cut the trim pieces for the cabinet (Fig. 6-22). These pieces can be made from ¾″ or wider quarter-round molding or picture framing.

They should be cut with beveled corners to make a frame that fits snugly inside the pine walls of the cabinet. Sand the cabinet and the trim pieces until smooth, and apply a stain as desired. The

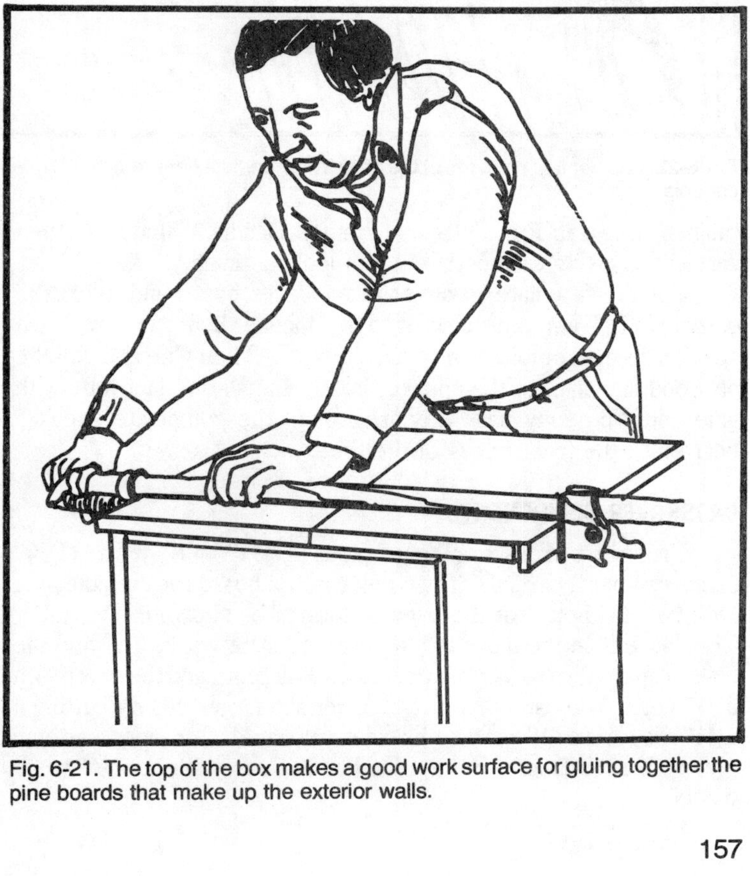

Fig. 6-21. The top of the box makes a good work surface for gluing together the pine boards that make up the exterior walls.

Fig. 6-22. Leather tab at the top of the grille frame permits easy removal to use controls.

cabinet shown in Fig. 6-16 was stained Danish Walnut and then varnished, giving a smooth and rich looking finish.

Cut the faceplate to cover the speaker board and make the cutouts in it. Temporarily nail it to the speaker board, and try each speaker in the combination to check the fit. Then glue and nail the plywood faceplate to the speaker board. Install the duct tube with glue and #6-penny nails driven through the composite speaker board into the front edges of the tube.

CROSSOVER COMPONENTS

Prepare a crossover board from a 7-¼″ × 8″ piece of ¼″ tempered hardboard. Drill two holes in the board for the controls. Drill two ⅜″ holes for the L-pads, located as shown in Fig. 6-23. Then assemble the parts for the crossover network. To wind the coils prepare the two forms, one with a 1-½″ core and the other with a ¾″ core. You can make the core for the larger coil by cutting a 1-½″ disc from ¾″ particle board or plywood. For sides use thin plywood or hardboard. For the smaller core, use a piece of ¾″ wood dowel.

The coils can be made on a temporary form; then the form is disassembled and the coil taped. You can make a break-apart form or a permanent form as you desire. For permanent forms, glue the side pieces to the core. Drill a hole in one side near the core. Measure out the correct length of wire; then start winding by feeding one end of the wire through the hole in the side of the form, from inside to outside. Allow about 6″ for a lead. Note that you must add about 12″ to the length of wire specified for a coil in order to allow for two leads. Start winding the coil, keeping the turns tight. It isn't necessary to count the turns. If you have access to a bridge, you can add a few feet to the specified length and then test the coil on the bridge. Remove enough wire to give the coil the exact

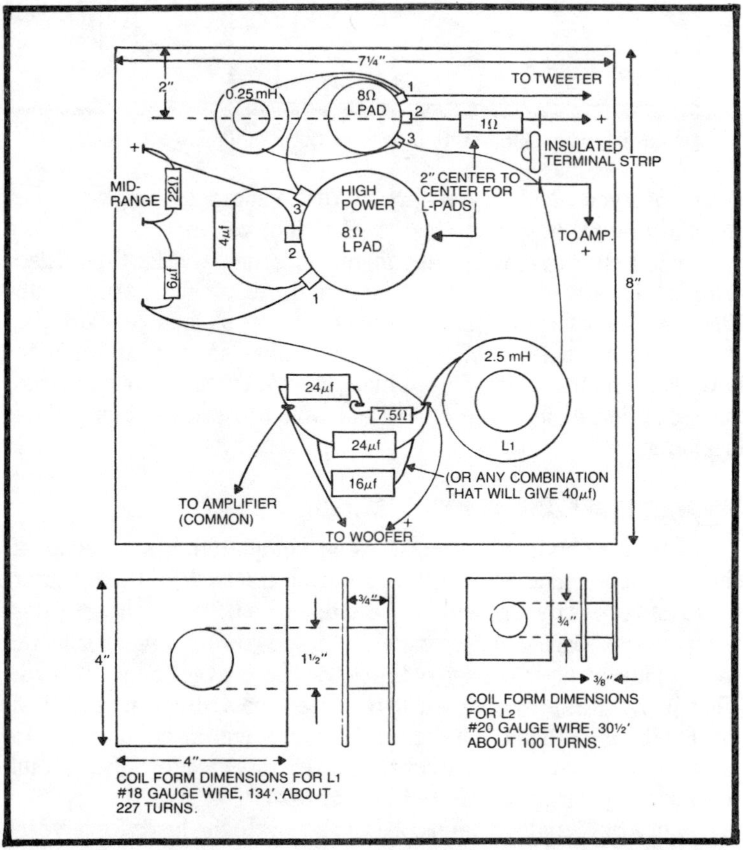

Fig. 6-23. Suggested crossover board layout and coil form dimensions for Project 14.

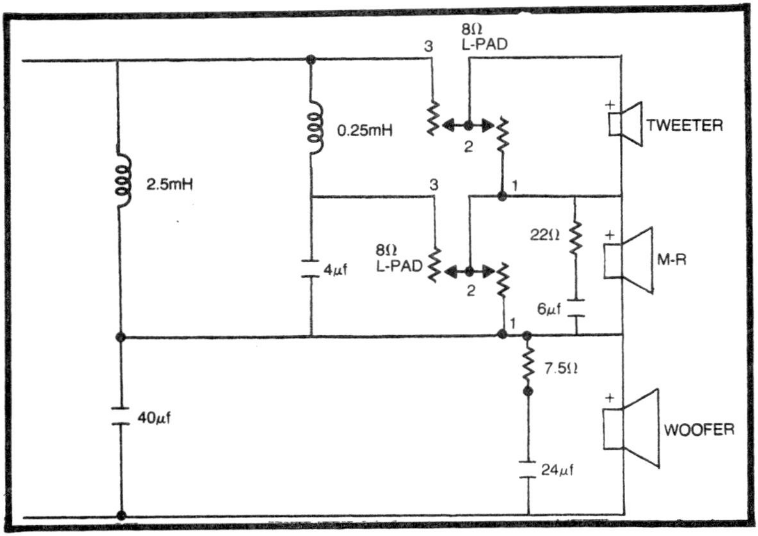

Fig. 6-24. Schematic diagram for crossover network for Project 14.

inductance you need. If you have no test equipment, just wind the coil and tape it. It will be close to the right value.

Install the crossover components according to Fig. 6-24. Also refer to Table 6-6. Mount the coils and the terminal strips with Elmer's Heavy Grip Cement or Leech's Heavy Adhesive. Wire the crossover and solder the connections. Give special attention to getting all the insulation off the magnet wire to make connections. Scrape, grind or heat the end of the wire to remove about ½″ of insulation.

SPEAKER INSTALLMENT AND TESTING

When the crossover network is completed, run a bead of silicone rubber glue around the cutout on the back of the speaker board and press the crossover board down on it. When the glue has set, turn the board over and install each speaker by running silicone rubber glue around the edge of the proper hole in the speaker board. Then press the speaker down on the glue. Leave the board for a half day for the glue to set. Then turn the board over again, use care to prevent damage to the speakers, and wire the speakers. Be careful to observe polarity as marked in Fig. 6-24.

Cut a 36″ length of lamp cord and separate the leads for several inches at each end. Strip ½″ of insulation from the leads. Feed the leads at one end of the cord through the holes in the back of the box.

Connect these leads to the lugs on a screw terminal strip, and mount the strip to the outside of the box with glue and tacks. Caulk the inner joints of the box and the inner holes for the terminals in the back. Line the inside of the box with 2″ to 3″ of damping material on the side walls, with a double layer behind the woofer. Run a strip of mortite or foam weather stripping material around the front edges of the particle board walls. Connect the lamp cord from the inside of the back to the proper points on the crossover network.

Before you install the speaker board, you should test the speaker. Set up the speaker board. Turn off your receiver, and turn the volume control down all the way. Connect the receiver to the speaker terminals at the back of the box. Turn the receiver on and turn up the volume just enough to easily hear the sound. Check to see if the bass comes from the woofer, the middles from the mid-range speaker, and the highs from the tweeter. Manipulate the controls to see if they are working properly. When you are satisfied that all is right, turn off the receiver and disconnect it. Install the speaker board with #8 × 2″ screws. If you have to remove the speaker board for any reason and find the screws getting loose, you can later switch to #9 screws. Check the woofer polarity by connecting a flashlight battery to the terminals on the back. When the cone moves forward on connection and to the rear on break, mark the terminal connected to the positive pole of the battery with a plus.

GRILLE FRAME

Make up a grille frame from 1″ × 2″ pine. Stretch your grille cloth across it and staple it to the front of the grille frame near the

Table 6-6. List of components for Project 14.

1	15″ woofer, Oaktron model Super Big Mag 15
1	5″ mid-range speaker, Pioneer, MStk1#PION-MID-10-8
1	1-¾″ tweeter, CTS Model 2TA3.
	Note: Speakers available from McGee Radio Co.
2	8 ohm L-pads, one preferably heavy duty (for Mid-range)
1	40 μf non-polarized electrolytic capacitor or any combination to make 40 μf.
1	4 μf non-polarized electrolytic capacitor
1	6 μf non-polarized electrolytic capacitor
1	24 μf non-polarized electrolytic capacitor
1	22 ohm 10 watt resistor
1	7.5 ohm 10 watt resistor
1 lb.	#18 gauge magnet wire
1/10 lb.	#20 gauge magnet wire.

edges. Then, after making sure that the grille cloth is stretched tight enough to hold, cover the staples by gluing the previously finished trim to the frame. The glue that goes between the trim and grille frame will help to hold the grille cloth.

You can install the grille frame by gluing strips of foam at several points around it to make a friction fit with the inner walls of the cabinet. Or you can use a cabinet door catch mounted on a recessed block at each side of the grille frame with a matching receptacle on the faceplate.

If you find the bass too much of a good thing, try moving the speakers farther from the room corners. Another ploy is to tune the box to a lower frequency by adding about 2-½" to the length of the bass reflex duct. You can experiment with the length and choose the length that gives the kind of bass you want. The only special equipment you need to get the most from this speaker is a room large enough to do justice to the low frequencies it can reproduce.

Equalized Speaker Systems

Although bass reflex speakers have often captured the imagination of speaker engineers and listeners alike, ported box systems have always required more careful design than the simple closed box. Another disadvantage for people who like their music loud is a flaw that many owners of reflex speakers don't consider. The woofer is essentially unbaffled for infrasonic bass signals. And these are the frequencies that require the maximum cone movement. Although such low bass contains almost no musical information, they are invariably present in the reproduction chain.

You can easily test any reflex speaker for this problem. Watch the cone. At high sound levels you can see it vibrate through great excursions. Or have someone turn an FM receiver with the volume control well up. You'll think the cone is going to jump out into your lap.

INFRASONIC FILTERS AND THEIR IMPORTANCE

One way to counter this tendency is to use a large woofer, such as the 15″ driver in the previous project. But it does seem too bad to limit the reflex to a few mammoth systems. There are a number of smaller drivers that work well in ported boxes if used with receivers that have infrasonic filters, such as the first two reflex projects in Chapter 6. But if the driver has too much compliance, one shudders at the thought of what can happen to it in a ported box. Even if it survives, the cone will undoubtedly distort the sound in its upper range because of excessive excursion at a simultaneous lower frequency. One such distortion is called *Doppler distortion*, similar to what happens when a car approaches you with the horn blaring. First the horn tone appears to rise in frequency. Then the tone is lowered as the car goes past you. When a speaker produces a steady upper tone and a lower frequency at the same time, the pumping action to reproduce the bass note can modulate the higher one.

If you examine your receiver and find no infrasonic control on it, that is no proof that none exists. On some models, the Advent 300 for example, the filter is built into the circuit. Evidently the engineers who designed that particular receiver consider such a filter mandatory. Only by comparing the performance of such a receiver with one that has no filter are you aware of the superiority of the receiver with the filter.

Considering the various options open to the speaker builder who wants to make a ported box system, it boils down to these choices:

- Use a large woofer, preferably 15" in diameter.
- Select a smaller woofer that is designed for reflex operation and use it in combination with an infrasonic filtered amplifier.
- Choose a small woofer with a really high compliance and build a closed box enclosure. Actually the builder who picks this woofer has another possibility.

ADVANTAGES OF AN EQUALIZER

When A.N. Thiele applied electrical filter analogy to the reflex problem, he developed 28 designs, or alignments. Of these, his first nine alignments work well with any good receiver that has an infrasonic filter. These produce a flat or nearly flat frequency response curve down to the bass cutoff frequency. These alignments have been widely used and acclaimed. But some of Thiele's other alignments, particularly the one he labeled #15, offer some tantalizing advantages.

If you build a reflex box and tune the system wrong, you will get a bump or a dip in the response curve. For example, if you tune the box too low, the bass will roll off at a higher than normal frequency. But if you add an auxiliary boost circuit to such a system, you will get an extended bass range with a sharp cutoff below the tuned frequency of the system. While a high compliance speaker can be used in a simple ported enclosure to get greater bass range, with some danger of overload at frequencies below the damping range of the port, the same speaker can yield even lower bass in a specially tuned box with an equalizer on the amplifier. And note this bonus. The equalizer can cut off the ultra-low bass so sharply that there is little chance of overload.

Moreover, the equalizer extends the bass by adding boost at the very frequencies where the speaker cone is highly damped by the box. The equalizer is usually tuned to a frequency just above that of the box resonance where the air in the port exerts maximum control on the woofer.

The design method, as originated by Thiele and refined by D. B. Keele, Jr., goes as follows: Use one of Thiele's first nine designs. Tune the box ½ octave below the normal tuning frequency and add an equalizer tuned to a frequency that is 7 percent above the box resonance. The design details are covered in TAB book No. 1064, *How To Design, Build, & Test Complete Speaker Systems*, if you are interested in designing your own system.

To summarize the advantages of an equalized reflex system over an unequalized reflex, here are three arguments for the equalizer:

— It reduces excessive cone motion and speaker distortion.
— It saves amplifier power and makes the amplifier sound cleaner.
— It extends the bass range of the system.

PROJECT 15: THE EQUALIZER

■ **Construction:** Use printed circuit card for easy building.
■ **Wiring:** Simple with card.
■ **Comments:** A necessity for reflex speakers and amplifiers or receivers that have no infrasonic filter. Also for specially designed systems such as Projects 16 and 17.

Here is a practical equalizer you can build and tune to any reflex speaker (Fig. 7-1). In fact you can use it with a few closed box speakers, but a closed box speaker should be designed for it, as is the one described later in this chapter. If you have a commercial reflex that you want to adapt to operation with the equalizer, you can use the reflex as it came from the factory. You might want to experiment with the box frequency. By lengthening the tuning duct, you can tune the enclosure to a slightly lower frequency and then apply the equalizer.

The circuit was designed by Pat Snyder of Speakerlab. Unlike Speakerlab's speaker system the equalizer isn't available as a kit, but you can order a circuit board from Speakerlab. The circuit board costs very little and is strongly recommended. With it you should get a working circuit at first wiring and with greater ease. Snyder's equalizer plugs into the amplifier or receiver by way of the tape monitor circuit. If you plan to build the equalizer, you should write to Speakerlab to get an update on any new developments in their circuit, such as might occur if the one specified here became unavailable. You'll probably want to get their publication, "Bass Reflex Equalizer," which costs very little. To get the current prices of this booklet and other such materials, write to Department BRE at Speakerlab and ask for a catalog. Their address is in the appendix.

Circuit Adjustments

Snyder's equalizer differs in at least one respect from the theoretically "correct" equalizer specified by the Thiele data. Thiele suggested an equalizer with a 6 dB peak at the bass boost frequency. Such a peak would theoretically flatten the response curve when the box is tuned ½ octave too low. Snyder made the height of the peak variable by using potentiometers to control the amount of boost. Although this method may seem less precise, it does offer much more flexibility in matching the equalizer to any speaker system.

The only other part of the circuit that you might need to adjust is the frequency of the bass boost. If you build one of the projects described in this chapter, the correct values are specified for C_2 and C_3. If you want to try the equalizer on another speaker, here is how to do it. Try to find out the tuned frequency of the box by using a test record, or an audio generator if one is available. Feed a strong signal to the speaker and watch the woofer cone carefully. At the port frequency the woofer cone will appear to stand still. Above and below that frequency the cone will move freely. When you find the box frequency, calculate the value of C_2 and C_3 by this formula:

$$C_x = C_3 = \frac{23.2}{\text{Box f}}$$

If the calculated value turns out to be one that isn't available, use the nearest smaller value. For example, if the frequency of the box is 32 Hz, the calculated value of the capacitors would be:

$$C_2 = C_3 = \frac{23.2}{32}$$

$$= 0.725\ \mu f$$

Fig. 7-1. The equalizer and the speaker system shown here are tuned to each other.

This isn't a commonly available value, but you can find capacitors with a value of 0.68 μf. That would give you a boost frequency of 35 Hz. Of course, if necessary, you can combine two or more capacitors in parallel to get the value you want. Since you will probably not be able to determine the precise frequency of the box without test equipment, the nearest value to that obtained by the formula above will be close enough for experimentation.

Using the Equalizer

Here is some advice from Pat Snyder on how to use the equalizer, "Take a bass reflex system that already performs well and hook it up to the filter. Or even just put a likely looking speaker in a fairly plausible box and adust the vent until it sounds best.

"Then hook up the bass reflex equalizer. For a start you might use $C_2 = C_3 = 0.5\ \mu$f so faux = 46 Hz. Run the peak up and down and see what happens. If this has no effect on the sound and faux is too low, raise it. If varying the peak height definitely affects the sound

Table 7-1. Parts list for the equalizer.

1—Circuit Card (optional)	2—150Ω, 5%, ¼W Carbon Film Resistor (Color code: Brown-Green-Brown-Gold)
1—117 V Primary, 24 or 25 VCT Secondary, 85mA (or more) Transformer.	4—1K, 5%, ½W Carbon Composition Resistor (Brown-Black-Red-Gold)
4—1N4002 or other 100 V. 1A(or greater) Rectifying Diodes.	2—47 K, 5%, ¼ W Carbon Film Resistor (Yellow-Purple-Orange-Gold).
4—1000 μf, 25 V or Greater Electrolytic Capacitors.	4—100 K, 5%, ¼ W Carbon Film Resistor (Brown-Black-Yellow-Gold).
1—LED (Light Emitting Diode).	2—.10μf, 25 V or Greater Capacitor
2—Raytheon 4739 Integrated Circuit Operational Amplifiers	2—.47 μf, 25 V or Greater Capacitor.
2—14-Pin Dual-in-Line IC Sockets (Optional).	4—Phono Jacks for Inputs and Outputs.
2—100Ω Trim Pots for Mounting Directly on Printed Circuit Board	1—On-Off Switch.
1—1K, 5%, ½W carbon comp. resistor goes next to LED	1—Fuseholder and 1 A Fuse.
	1—Line Cord and Plug.

"Plus capacitors for C_2 & C_3. These to be selected to match speaker system. See text."

but does nothing to *improve* it, faux is probably to high. Reduce it. If large peak heights make the bass overstrong but solid, you're in the ballpark. Reduce the peak height to where you get maximum realistic bass. When you get it set right, bass will go low and be solid, and visible cone motion will be substantially reduced over what it is without the filter.

"This approach will not give an 'optimum' design, merely one that is very good. Each of Thiele's alignments is optimum in the sense that each satisfies some different mathematical criterion. A trial-and-error design will satisfy only one criterion—excellent performance."

BUILDING THE EQUALIZER

The first step in building the equalizer is to obtain the parts so that you will know how to drill the printed circuit card for each part (Table 7-1). If you live in a small city, you may have to order some parts, such as the transformers. Check the current publications from Speakerlab; they sometimes sell a few of the parts to go with the card.

Drilling Holes in the Circuit Card

When you have the parts assembled, you should make the location and size of the needed holes on the card (Fig. 7-2A). Check the diameter of the holes needed for the printed circuit jacks. These are sometimes designed to fit ¼″ holes, sometimes ⅜″ holes. The position of the jacks is marked on your card. Drill the holes, preferably using a small drill bit to locate the hole, and then the full size bit.

Use great care while drilling the board; hold it firmly against a block of wood. If you have access to a drill press, the job goes much easier. But you must take care to avoid damage to the circuit board.

The holes already in the card for a transformer are spaced too far apart for the Stancor P-8394 or the Triad F-115X transformers. You can use the prepared hole nearest the edge of the card and drill a new ⅛″ hole at 1-¾″ distance from it.

Next, drill new holes at the locations shown in Fig. 7-2A for the SPDT switch, the fuse holder, the ac cord, and the primary leads of the transformer. When you complete the hole drilling, install legs at the four corners of the equalizer (Fig. 7-3). You can use ⅛″ bolts 1-½″ long for these legs if you run a nut up each bolt and tighten it against the card. Next, install the four jacks (Fig. 7-2B). Run short lengths, about 1″ of wire from each jack to the nearest point on the underside of the card marked "in" or "out". Run insulated wire from the ground lug on the left output jack to the ground lug on the left input jack. Run a short length of wire from there to the ground lug on the right output. Then run another length of wire from the right output to the ground lug on the right input. Finally run a short length, about 1-½″, from right input ground to the point on the

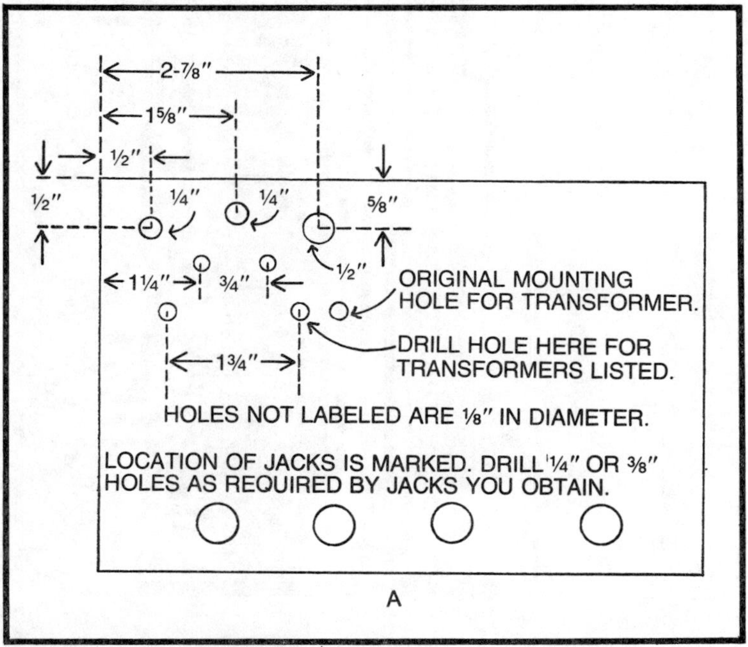

Fig. 7-2. (A). Drilling pattern in circuit card.

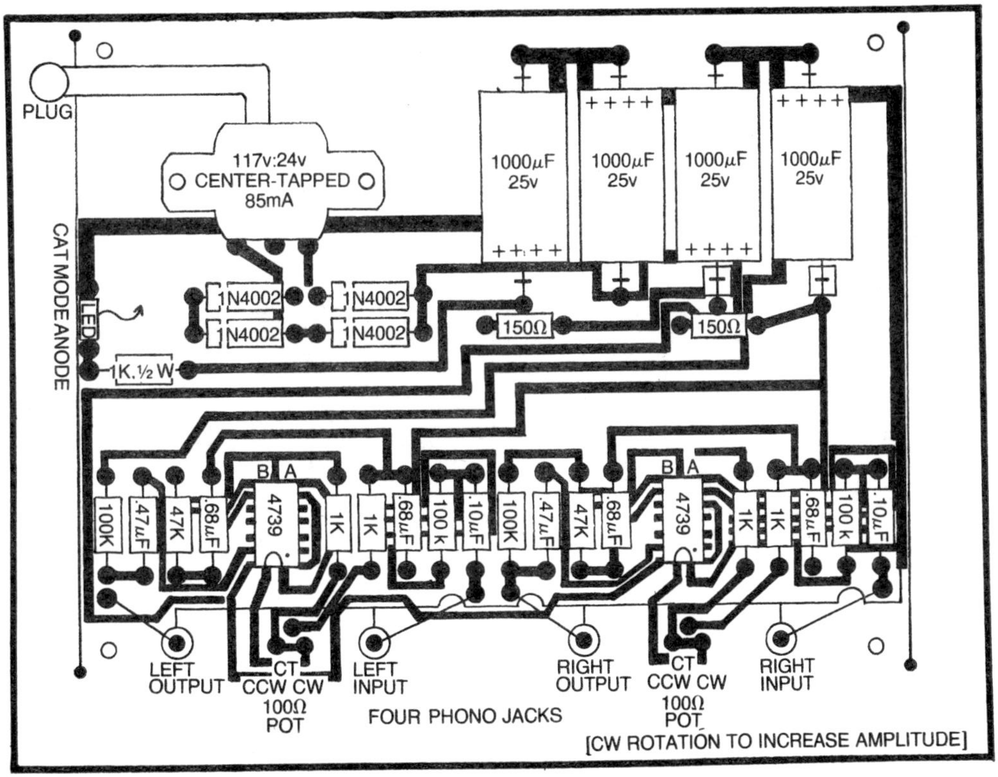

Fig. 7-2. (B). Parts placement on circuit card.

board marked "gnd." Check these wires to see that all are insulated from the hot terminals of the jacks. Solder all connections with rosin core solder.

Parts Installment

Install the IC sockets. Note that pin 1 on these sockets must be placed at the end nearest the jacks; this is the end with the half-moon indent. Use an iron with a small tip to solder the connections under the chassis, and use solder sparingly for the closely placed pins. Solder every pin. Inspect your work carefully with a magnifier to make sure you have no shorts.

Install the resistors next. They are small and pose no problem of access to other parts. If you should put in the capacitors first, though, you might have trouble reaching the resistors. That depends on the capacitors you buy. Push the leads down through the proper holes in the card. Then bend the lead on the underside to hold the part in place. After you have installed a half dozen or so, stop and solder the leads; then clip off the excess wire. Check your solder joints. They should be bright and smooth.

Install the other parts such as capacitors, diodes and the light-emitting diode (LED). Note that the value of the capacitors you install for C_2 and C_3 in each channel will depend on the speaker

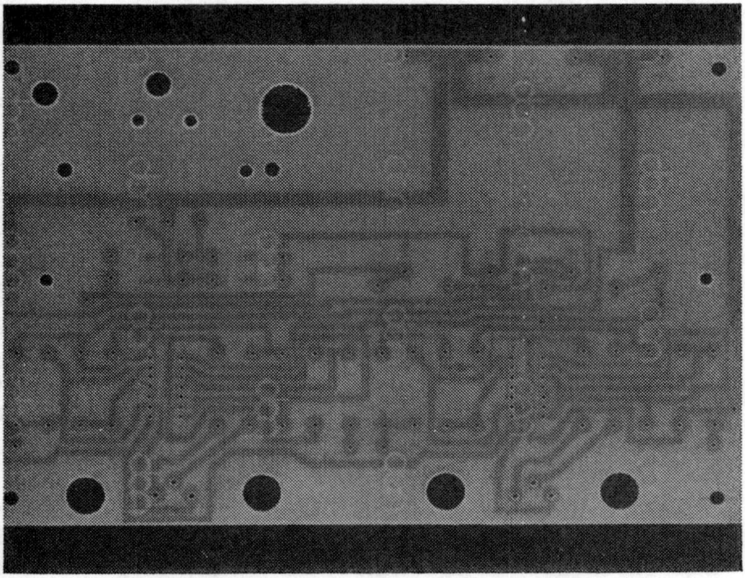

Fig. 7-3. Appearance of the equalizer circuit card after drilling.

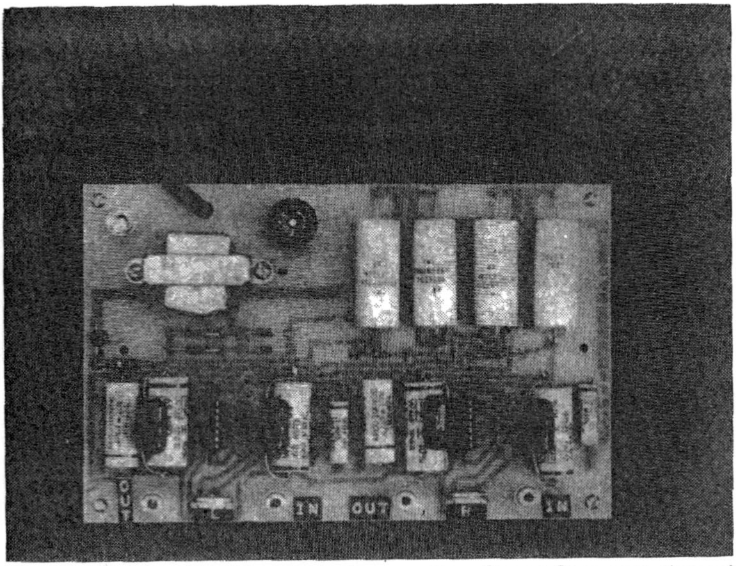

Fig. 7-4. The completed equalizer with capacitors C_2 and C_3 in each channel selected to match boost frequency to Project 16.

system you intend to use with the equalizer (Fig. 7-4). Give particular attention to the polarity of the electrolytic capacitors, the diodes and the LED. The electrolytics are plainly marked with a (+) at one end and a (−) at the other. The diodes have a band at one end, and the LED may have the cathode lead made shorter. It may have an indent on the body at the side of the cathode lead. Check the bubble pack for information before you discard it.

Set the transformer temporarily in place with the three secondary leads facing the side of the chassis with the jacks. Check to see how long to make these leads and then cut them off. Remove ¼″ of insulation from each lead and install the transformer, feeding the secondary leads through the three holes on that side of the transformer and the two primary leads through the holes you drilled on the other side. Note that the center tap on the secondary, the wire with the different color coding to the other two, goes through the middle hole. This center tap is yellow on the Stancor transformer, yellow striped on the Triad. Solder the secondary leads and clip off any excess wire.

Install the fuse holder and the switch. Run one of the primary leads from the transformer to the middle lug on the switch. Run the other one to one of the lugs on the remaining lug on the fuse. Solder all connections. Insert a 1 amp fuse in the fuseholder.

Go back and inspect your work. Use a magnifier to check all solder connections under the card. If everything looks okay you are ready to test. Advance the two controls on the equalizer card to full bass boost. Plug the equalizer into the tape monitor circuit of your receiver and turn the receiver and equalizer on. Try your receiver at low volume with the tape monitor circuit "out." When you switch it to "in," you should hear a difference in the sound unless you have tuned the equalizer frequency too low for the speaker system you are using. If you are using two capacitors to get the value you planned, try removing the capacitor of small value.

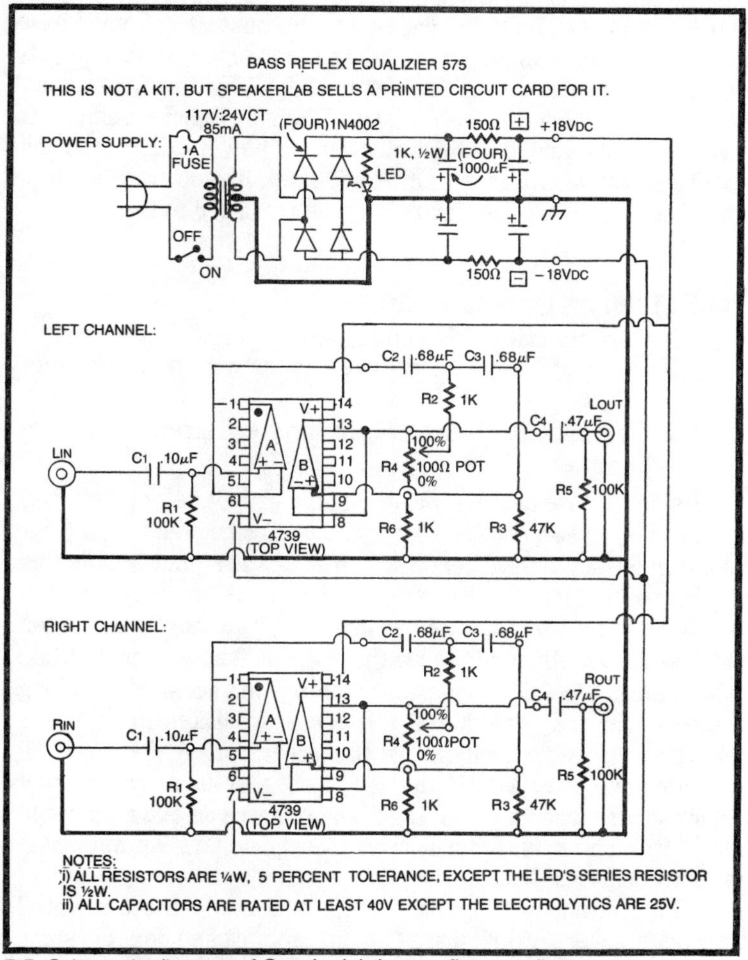

7-5. Schematic diagram of Speakerlab bass reflex equalizer.

If you still notice no change, you must do some trouble shooting. Speakerlab's Technical Compendium #3 on the Bass Reflex Equalizer lists some steps to take if you don't get a signal. But the chances are you will have no trouble unless you've made a mistake in wiring. If the thing doesn't work the first time, go back and inspect your wiring and the circuit diagram (Fig. 7-5). Have someone help you check it. I have built two of these gadgets and each one worked the first try, a better record than I've had from many kits. The first one more than paid for itself by resurrecting an old amplifier that I had discarded. I had decided that the amp just wasn't clean enough to justify using it with the rest of my stereo system. But then I hooked up the equalizer to it because of a phobia I have about testing new gadgets with my best equipment (Fig. 7-6). My daughter put on a record of "Song of Joy" by the Captain and Tennille. I couldn't believe my ears. The old amplifier seemed to have suddenly gained a hundred watts in power. Of course this particular amplifier had no infrasonic filter. If you have such an amplifier or receiver, get started on an equalizer today.

PROJECT 16: AN EQUALIZED REFLEX

- **Construction:** Use power saw.
- **Wiring:** Simple. Pre-wired crossover network (Fig. 7-10B).
- **Comments:** Deep bass. Requires equalizer. 4 ohms impedance.

As the picture of this speaker and the gadget before it shows, this speaker and an equalizer go together like ham and eggs (Fig. 7-1). A good ham-on-rye needs no help from the henhouse, but this speaker without this equalizer could be in trouble. Instead of the 0.6 to 0.7 cubic foot box it was designed for, the woofer has been installed in a box about 2-½ times that volume. And it's a ported one at that. Small acoustic suspension woofers don't ordinarily last long in boxes that are ported. But the equalizer makes the speaker behave as if it were designed for the larger ported box.

Someone may ask, "Why put a speaker in a system that requires more than twice the space and requires an extra gadget to boot?" How about this for an answer? Most listeners can appreciate an extra octave of bass.

This system was compared with one in which the identical components were installed in a 0.7 cubic foot closed box. On some program sources the difference was barely noticeable, more a

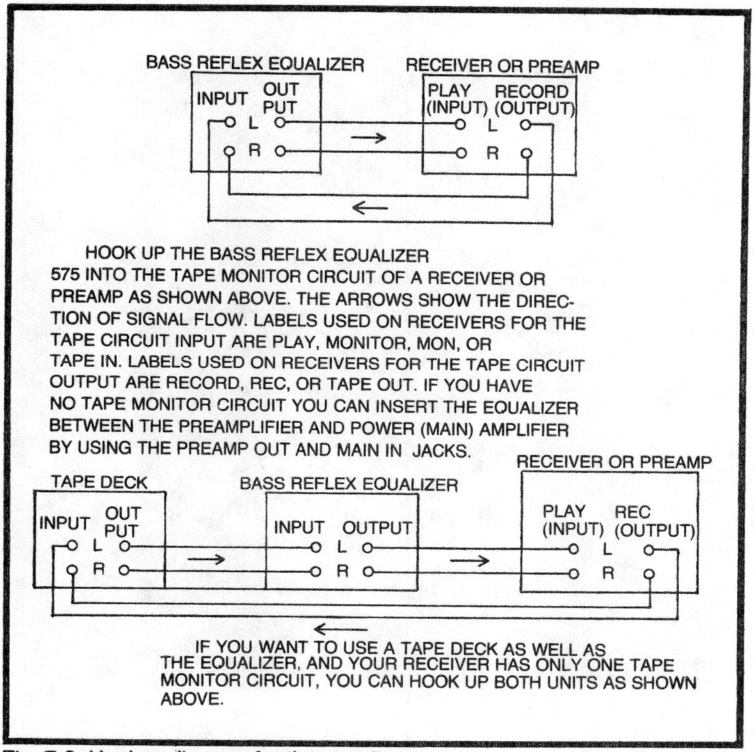

Fig. 7-6. Hookup diagram for the equalizer.

different quality of bass. The reflex bass was considered slightly more free and easy. But when a program source with extended bass was supplied, the difference was dramatic. A typical comment was that the larger equalized speaker gave much more exciting sound. In fact this equalized system has exceptional bass response for the size of the system. There is a limit to how much power at low frequencies you can expect from an 8″ cone, but within that limit this system delivers a deep true bass.

CONSTRUCTION PROCEDURE

The enclosure shown was made from ¾″ birch with a ½″ particle board liner (Figs. 7-8 and 7-9). The plywood was of fairly good quality, but not the equal of that used in some other projects in this book. That is why the particle board liner was used to brace the walls. You can substitute ¾″ particle board and use butt joints. Cover the particle board with some kind of veneer. If you do that,

Fig. 7-7. Project 16 with grille cloth installed.

just make sure you keep the internal dimensions equal to those shown in Fig. 7-10A so that the internal volume will be right. See Table 7-2 for materials.

Enclosure Work

Cut out the sides, top and bottom to the proper width. Then make the 45 degree cuts in the ends of the panels. Apply glue sparingly to the beveled end and assemble the enclosure shell on your work bench, using straps or ropes to hold the parts together under pressure while the glue sets.

Measure the exact internal width of your shell and cut particle board pieces to fit inside the top and bottom of the shell. Set these pieces in place and measure the exact length of the side pieces to fit inside the top and bottom pieces. Cut the side pieces. Install the particle board liner by gluing and inserting the top and bottom pieces first and then the side pieces. The sides will hold the other pieces tightly against the top and bottom, but you should use pieces of

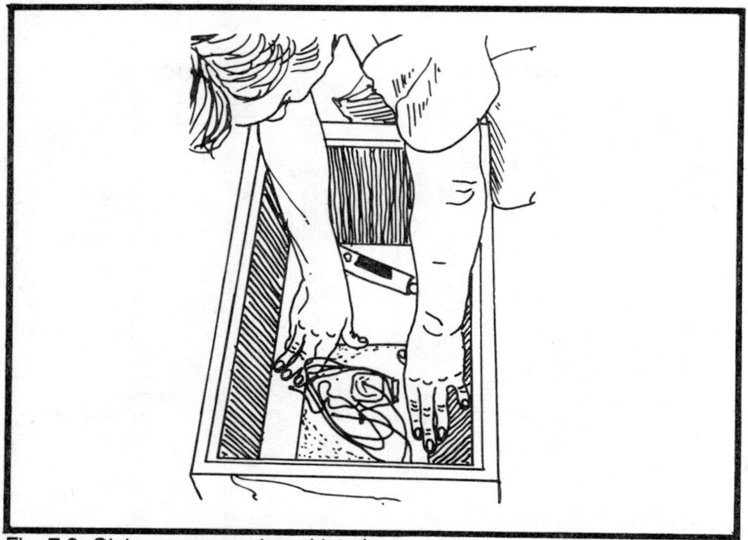

Fig. 7-8. Gluing crossover board into box.

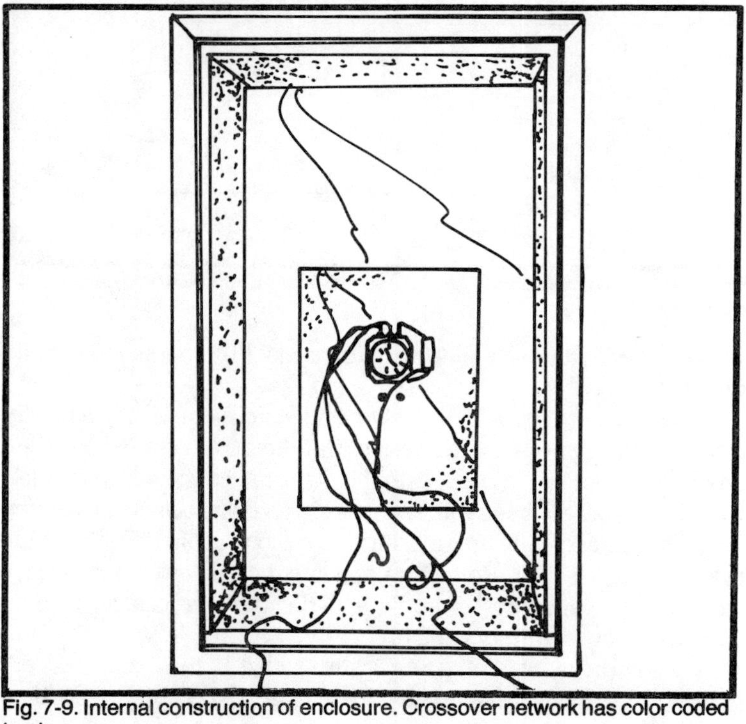

Fig. 7-9. Internal construction of enclosure. Crossover network has color coded leads.

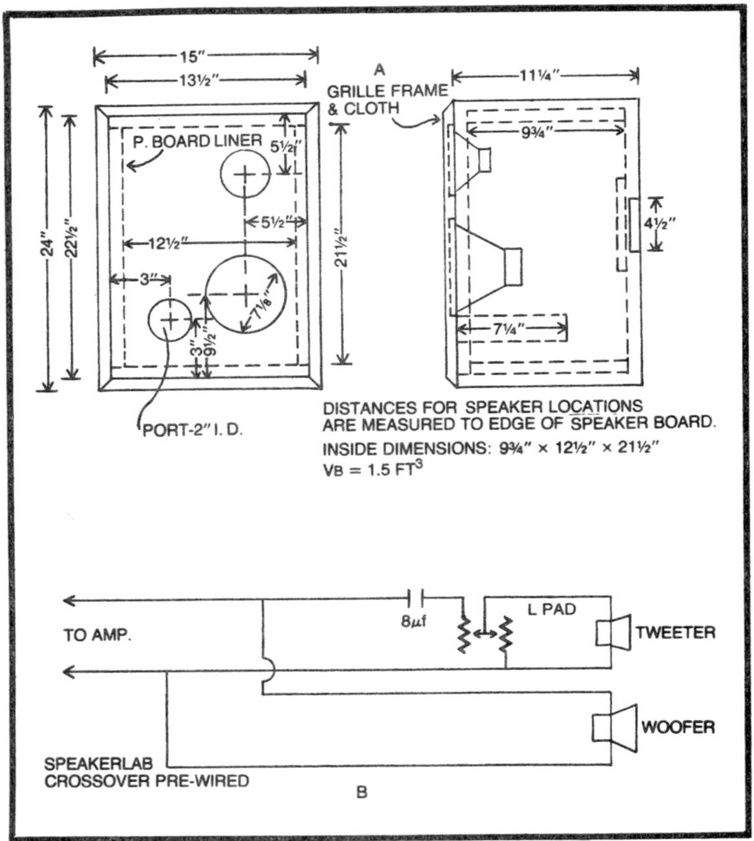

Fig. 7-10. (A). Construction plans for Project 16. (B). A pre-wired Speakerlab crossover.

wood to wedge the side pieces tightly against the outer walls until the glue sets.

Sand the outer walls of the box and stain them. Apply a finish coat of varnish or oils as you desire. Cut the back from ¾″ particle board. Make a 4-½″ × 5-½″ cutout in it for the crossover network. Glue and install the back with heavy weights on it to apply pressure while the glue sets. Then run a bead of silicone rubber glue around the cutout in the back. Press the crossover board down on the glue, centering the speaker terminals and the tweeter control in the cutout. Caulk the interior joints.

Cover the interior of the enclosure with a 1″ to 2″ thickness of polyester batting or fiber glass batting (Fig. 7-11). Thread the speaker leads out through the material you put on the back.

Speaker Board Directions

Cut the speaker board from ¾″ particle board and make the cutouts for the speakers and port. Note that the port is a tube with a 2″ internal dimeter. If you don't have a tube of this diameter, you can use a port with a square cross-section. Just make the internal dimensions 1-¾″ × 1-¾″ and make sure the depth of the port is 7¼″, from the front of the speaker board to the rear edge of the tube. Paint the speaker board flat black, if you desire.

Set the cabinet on its back, propped up so you can reach the terminals on the back. Run a bead of silicone rubber glue around the upper edge of the particle board liner (Fig. 7-12). Set the speaker board down on the glue, feeding the speaker leads out through the proper holes so they won't get in the way of the board. Press the board down hard.

Set the speakers on the board and locate the proper leads for each one: yellow for the tweeter's positive terminal, yellow/black for its negative, blue for the woofer's positive terminal and blue/black for its negative terminal. Push the proper connector onto each terminal and set the speakers in their holes. Turn off your receiver and turn the volume all the way down. Connect the receiver to the

Table 7-2. Materials list for Project 16.

¾″ Birch plywood, pieces needed:		
2	11-¼″ × 15″	Top & bottom
2	11-¼″ × 24″	Sides
1 × 12 pine boards can be substituted for plywood if particle board liner is used.		
¾″ Particle board, pieces needed:		
2	13-½″ × 22-½″	Speaker board & back
½″ Particle board, pieces needed:		
2	9-¾″ × 13-½″	Top & bottom
2	9-¾″ × 21-½″	Sides
¾″ Particle board or plywood, one piece.		
1	15″ × 24″	Grille frame, cut out to make a frame, 2″ wide.
2″ I.D. Tube 7¼″ long		Vent tube.
Speakers and components:		
1	S1s kit, Speakerlab, Specify woofer W804R. This kit includes woofer, tweeter and crossover.	
4	0.68μf 25 v., or greater, capacitors	
4	0.22μf 25 v., or greater, capacitors	
	These capacitors are used to make up four 0.9 μf capacitors for C_2 and C_3 in each channel of equalizer. See photo of equalizer.	

Fig. 7-11. Line interior walls with damping material before installing speaker board.

terminals on the back of the enclosure. Turn the receiver on and bring up the volume just enough to hear the sound clearly. Check to see that the bass comes from the woofer and the treble from the tweeter and it is controllable by the L-pad. Don't expect much bass from the unbaffled woofer. If the speakers operate properly, turn off the receiver and disconnect the speaker. Remove the speakers from the holes and run a bead of silicone rubber glue around each speaker cutout. Set the speakers down on the board and rotate them slightly to spread the glue. Press them down hard, and leave the cabinet in that position for a half day or more.

TUNING THE EQUALIZER

For optimum performance the equalizer and speaker should be tuned to work together. Theoretically the equalizer should be tuned to a frequency about 7 percent above that of the reflex enclosure. Here the enclosure is tuned to 25 Hz, so the equalizer should be tuned to about 27 Hz. The capacitors chosen for C_2 and C_3 of each channel consisted of two individual capacitors wired in parallel. The value of the capacitors was 0.68 μf and 0.22 μf. The tolerance of the components was 10 percent. When they were wired into the equalizer, the peak frequencies on each channel were within 1 Hz of each other and right on target. The performance was so good that no other substitutions were tried.

If you like to experiment, you can shorten the tube to raise the frequency of the box, but you will probably also need to raise the tuning frequency of the equalizer. This change might give you more usable bass, but would very likely produce a bump in the response curve that would require you to roll back the boost on the equalizer. The chief advantage of such a change would be to use the speaker at high sound intensity levels where the higher frequency roll-off in the

Fig. 7-12. Run a bead of silicone rubber glue around the upper edge of the particle board liner before pressing down speaker board.

equalizer might permit greater output. If you want to try such an experiment, you could shorten the tube to about 3-½″ long. This would probably raise the box frequency to about 30-33 Hz. Then remove the 0.22 μf capacitors from the equalizer tuning capacitors, leaving only the 0.68 capacitors in the circuit. This would raise the equalizer boost frequency to 35 Hz.

Experiment if you want. If you just build up the equalizer, tune it with the 0.9 μf total capacitance as specified, and build the box according to Fig. 7-10A, this system should give you exceptionally wide range sound for its size.

PROJECT 17: AN EQUALIZED CLOSED BOX SYSTEM

- **Construction:** Easy.
- **Wiring:** simple. Pre-wired crossover.
- **Comments:** Deep bass. Requires equalizer. 4 ohms impedance.

The bass reflex equalizer was obviously designed for use with bass reflexes, but nobody will make a federal case if you use one with a closed box (Fig. 7-13). You wouldn't like the sound you would get with some combinations of equalizers and boxes, and in other mismatched sets you'd find that the equalizer made no apparent difference in the performance of the speaker. If a closed box speaker is to work well with an equalizer, the two must be designed for each other.

This brings up a logical question. Why tamper with a simple, virtually foolproof system such as a closed box speaker? Again, as with an acoustic suspension woofer that is converted to equalized reflex operation, the reason for equalizing a closed box system is to get an extended bass response. But this can't be done properly by just hooking up an equalizer to any speaker system. Most closed box speakers have a slight hump in the response curve at resonance, and if more bass boost were added, the hump could get out of hand. The hump exists because most closed box speakers are designed to have a system Q of at least 1 or greater. The term Q, as used here, denotes the degree of resonance magnification in the system. A magnification of 1 signifies that the speaker is not overdamped. If you decrease the size of the box by filling some of the space, you would produce a system with a higher Q and less damping. If the Q is high enough, you would get a noticeable bass boom. Suppose you put the same effect on the woofer as increasing the compliance of its suspension system. The cone would move more freely and could be undetectable because in a system of lower

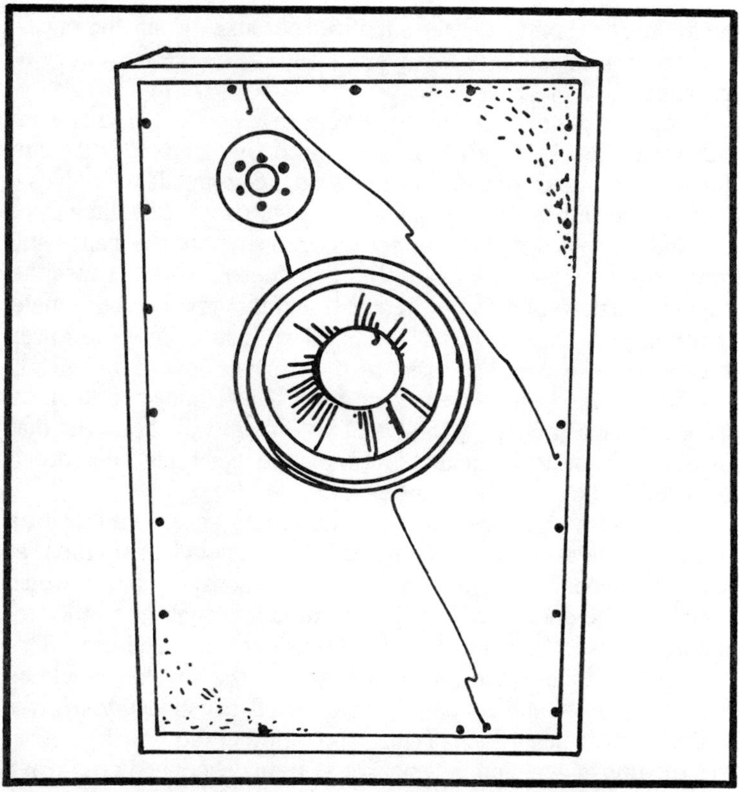

Fig. 7-13. Drivers used in this equalized closed box speaker are the same ones specified for the more compact enclosure of Project 7.

Q, the low end response would fall off at a smoother rate with no humps. While this would produce an inherently flatter response curve, the bass droop would start at a much higher frequency. At a Q of 0.7 the system would have an essentially flat response down to cutoff. With so-called "critical damping," a Q of 0.5, the response would be down 6 dB at resonance.

CLOSED BOX DESIGN APPROACH

The traditional approach to closed box design is to aim at a system of Q of about 1. This figure is recognized as a good compomise between good damping and good bass range. A lower Q sacrifices bass, and a higher Q can cause noticeable boom. An exception is made on some speakers that are designed with a higher Q on the theory that rock fans like a punchy bass response. Here again some bass range may be sacrificed because the smaller enclosure that raises the Q of the system also raises the frequency of resonance.

Some critics of closed box speakers have pointed out that even with a final Q of 1, the system is not perfectly damped. And some closed box designers aim for a Q of 0.7 on the grounds that it gives a flatter curve with better damping. People who want to eat their cake and have it, too, would like to set the frequency of resonance still lower to extend the bass range. But that doesn't work because the magnet starts to overdamp the cone, and this produces the mentioned bass droop. It would be nice if we could have the lower frequency of resonance and a flat response curve down to that point.

Enter the bass reflex equalizer. Thiele specified that an equalizer for alignment #15 should provide a 6 dB boost. If that figure isn't familiar, it should be. That is the exact degree of droop for a closed box speaker system with a Q of 0.5.

So, and here is the theory behind using an equalizer with a closed box speaker, why not enlarge the box enough to set the Q at 0.5? This would lower the frequency of resonance, but it would overdamp the speaker. But why not tune an equalizer to the frequency of resonance and add a 6 dB boost at that frequency? This combination should give the advantage that the speaker would be critically damped and yet should produce a flat response to a lower frequency than in any kind of closed box without the equalizer. That was the thinking behind the speaker system described here. When compared to Project 7, a smaller closed box system with the same components, the equalized speaker had no difficulty in proving its deeper bass. Otherwise the sound was very similar. This system has the disadvantage of occupying more space and requiring an equalizer. But if deep bass is what you want, an equalized closed box system is worth a try. If you want to apply the technique to another closed box system, you can find design information in TAB book No. 1064, *How To Design, Build, and Test Complete Speaker Systems*.

CONSTRUCTION DETAILS

The enclosure shown was designed to be covered with an inexpensive plastic veneer (Fig. 7-14). You can substitute hardwood plywood for a more elegant box, but the ¾" particle board should be retained for the back and speaker board. The construction procedure here will describe that used for the enclosure in Fig. 7-14. You can change dimensions to match your needs, but don't

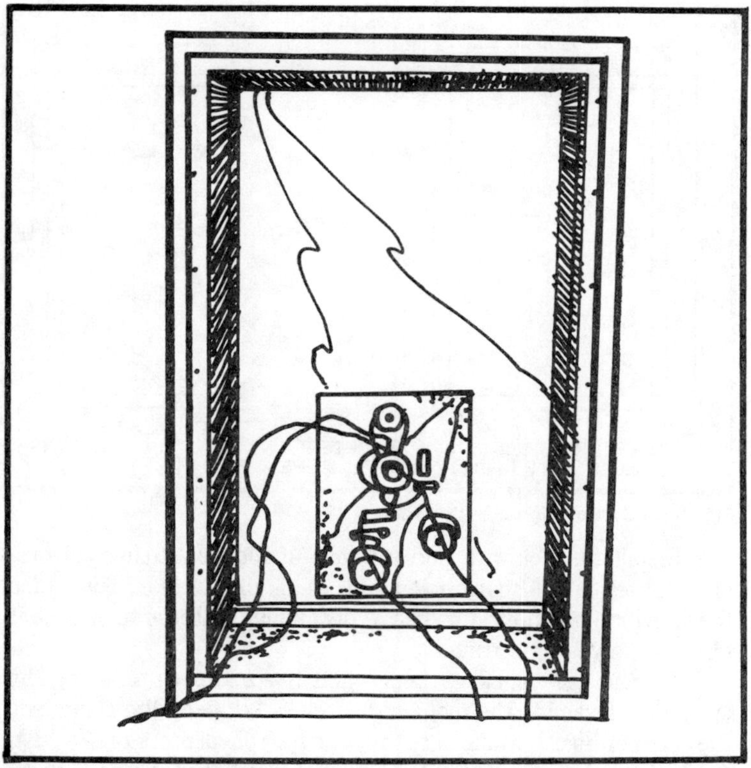

Fig. 7-14. Internal construction of enclosure. Plans show higher crossover mounting to give longer leads to speakers.

significantly alter the internal cubic volume. It should be about 3 cubic feet. See Fig. 7-15 and Table 7-3.

Cut out the sides, top and bottom pieces. Glue and nail the bottom to the sides and then the top. Measure the exact internal width of the box and cut four lengths of ¾″ × ¾″ material for cleats at the top and bottom. Install these cleats with glue and nails, ¾″ from the front and back edges of top and bottom. Cut side cleats to fit and install them. Then cut glue blocks to fit each corner in the space between the front and rear cleats. Install these corner blocks with glue.

Cut out the back and make a 5-½″ × 4-½″ cutout in the center of the back. You may notice that one picture shows the crossover network located near the bottom of the enclosure (Fig. 7-14). Either location will do but the center position provides longer leads for easier speaker wiring (Fig. 7-16).

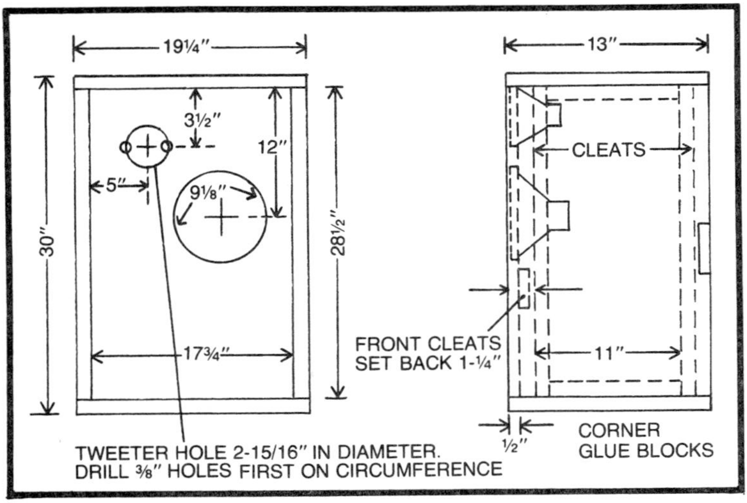

Fig. 7-15. Construction plans for Project 17.

Install the back with glue and nails. If you want to use screws, don't forget to drill shank holes first. Caulk the interior joints. Line the interior of the box with 2" to 3" of damping material such as fiber glass or polyester batting.

Prepare the speaker board and make the cutouts for the speakers. Examine the matching surfaces between the cleats and the speaker board. Make sure they are free of splinters of sawdust. Apply a surplus of wood glue and install the board. Drive it down with nails or screws.

Before you apply the veneer, you can paint the front flat black if you desire. The cabient will require at least 8-½' of veneer material but you should figure about 3 yards. Cut a strip about 14" wide and start on the bottom of the box. Align one edge of the veneer with the

Table 7-3. Materials list for Project 17.

Enclosure Materials	Speakers and Components Needed
¾" Particle board, pieces needed: (Industrial grade board) 2 13" × 19-¼" Top & bottom 2 13" × 28-½" Sides 2 17-¾" × 28-½" Speaker board & back. ¾" × ¾" pine, pieces needed: 4 17-¾" long Cleats, top & bottom. 4 27" long Cleats, sides. 4 9-½" long Corner glue blocks. Plastic veneer material. 1 14" × 100" Enclosure covering. ⅜" plywood 1 12-15/16" × 28-7/16" Grille frame, cut out to fit over speakers.	1 S2s kit from Speakerlab, including 10" woofer, tweeter, and crossover network. 4 0.47 μf @ 25 v, or greater, capacitors 4 0.047μf @ 25v., or greater, capacitors These capacitors are to be used in parallel pairs to make four capacitors (C_2 & C_3, each channel) of 0.517 μf to tune the equalizer to the speaker

back edge of the enclosure. Make sure the material runs straight with the box walls. If you find it getting off course, pull it loose and start over again.

When you get around the box and reach the starting point on the bottom, overlap it and cut off the excess length. Notch the corners and fold the veneer over to look like a beveled joint.

Prepare the grille board and stretch the grille cloth over it, stapling it at the back. The grille board, with the cloth installed, should fit inside the enclosure walls tightly enough for a friction fit. If you use a very thin grille cloth, you should allow about 1/16″ difference in grille board dimensions, as compared to those of the speaker board, to accommodate the grille cloth. For heavier cloth, allow ⅛″ difference.

You'll find this system gives extended bass and a feeling of much deeper sound. Its only disadvantage comes if you try to pump

Fig. 7-16. Speaker wiring in the center position is easier because of longer leads.

too much bass boost into it at very high power levels. A 10″ cone can produce excellent bass response but not at unlimited power output.

In judging whether you should build this system, you might consider how loud you like your music. If you like clean wide range sound at average loud levels, this a good choice. But if you must test the endurance of your equipment, and your neighbors, you might better put the same speakers in a box half this size. In that case you should build Project 7. It's a question that only you can answer.

Omnidirectional Speakers

One thing that can be said for omnidirectional speakers is they sound different. While the difference may not be startling, *omni* speakers give an effect of greater depth that many listeners find irresistable. Critics say they make single instruments or vocalists sound too wide and that precise location of instruments in space is impossible because of spatial smear. But most listeners agree that with massed voices or other large groups, they sound great. One argument in favor of omni speakers is that at a live concert you hear mostly reflected sound. Omni speakers, their fans say, add the reflections that make music more natural. Also, they add, the treble range is more widely dispersed with an omni speaker.

There is only one way to decide whether to go omni or not, and that is to listen to some omni speakers. For those who want the distinctive omni sound, here are two projects that were applauded by those who heard them. The first one offers a lot of sound for the money and is easy to build. The second will challenge your cabinet-making ability and costs considerably more, with much of its extra cost going into wood for the fine piece of furniture it is. Look them over and see if one will fit your needs and your listening room.

PROJECT 18: THE TOWER

- **Construction:** Fairly easy.
- **Wiring:** Simple crossover, but needs care with multiple tweeters.
- **Comments:** Good range for small floor space required. Typical omni spread.

Fig. 8-1. Tower enclosure with lineup of drivers used in the system.

Here is a speaker with a wrap around grille cloth that can hide minor mistakes in carpentry (Fig. 8-1). The box material is ½″ particle board joined with simple butt joints. The ½″ thickness is adequate in this case because of the narrow side panels.

The unusual shape poses no problems because a partition in the middle divides it into two equal enclosures. The speaker lineup consists of two 8″ woofers, four 3-½″ mid-tweeters, and a Motorola piezo-electric super tweeter per enclosure. The woofers have 16-ohm voice coils, so two of them in parallel show an impedance of 8 ohms.

In spite of what you may suspect by the number of drivers, this is a low-cost system. With inflation changing prices, it's risky to quote prices. A set of seven drivers will probably cost less than two drivers in several other projects. By spreading the mid-range and treble around the room, instead of concentrating it, no mid-range or

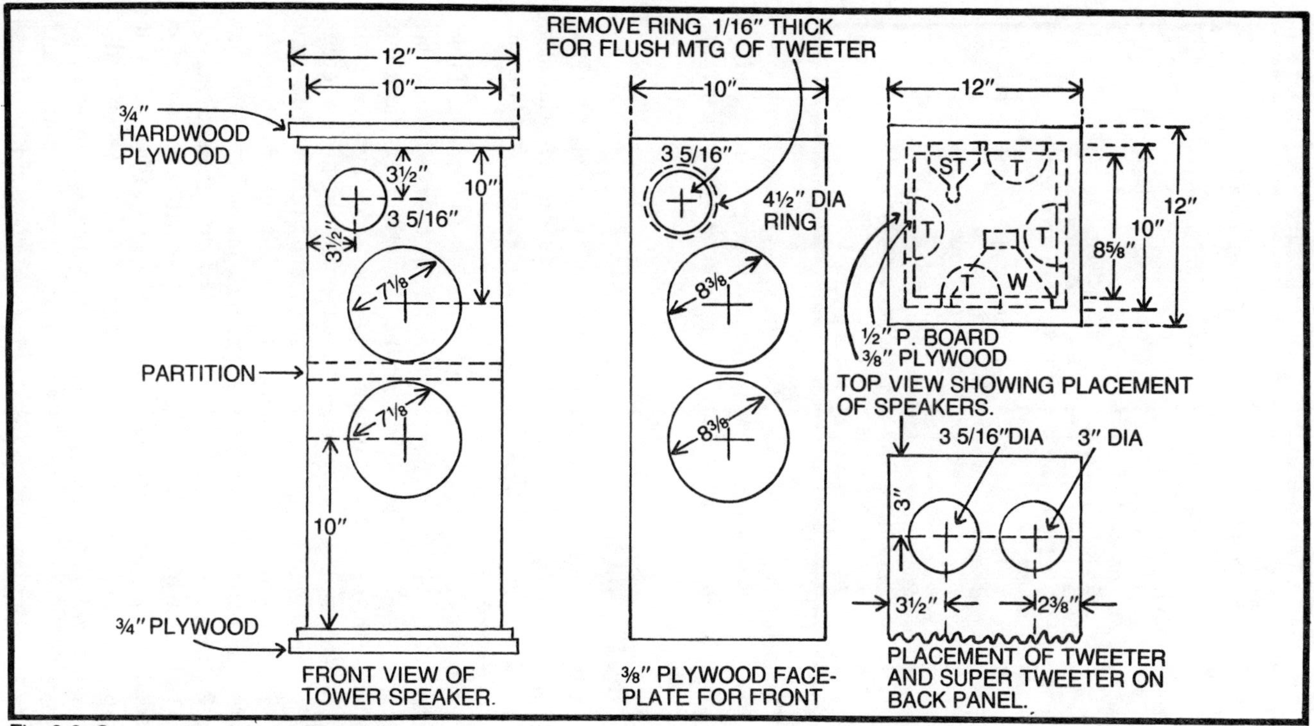

Fig. 8-2. Construction plans for Project 18.

Table 8-1. Materials list for Project 18.

½" particle board, pieces needed:		
2	10" × 28-½"	Woofer board & back.
2	8-⅝" × 28-½"	Sides
2	8-⅝" × 9"	Partition and liner for bottom.
⅜" plywood:		
1	10" × 28-½"	Faceplate for woofer board
¾" Hardwood plywood, pieces needed:		
2	12" × 12"	Top & bottom
¾" Wood veneer stripping, to match top & bottom:		
1	8' roll per cabinet.	To cover plywood edges.
½" Quarterround, pieces needed:		
8	approx. 11" long	To cover edges of grille cloth.
Speakers and components needed:		
2	8" woofers, 16 ohm impedance, Model ROLA-8	
4	3-½" mid-tweeters, CTS Model MT3X	
1	Super tweeter, Motorola 59067 Super Horn.	
1	16 μf, non-polarized electrolytic capacitor.	
1	4 μf, non-polarized electrolytic capacitor	
1	7.5 ohm resistor, 2 watts or greater.	

tweeter controls are necessary. And the crossover network consists of three inexpensive components. So if you have wanted an omni speaker but worried about the cost, here's your chance to build one.

CUTTING INSTRUCTIONS

Cut out the sides. Note that the front and back panels are wider than the side panels to permit a square cross-section with butt joints. Also, there is an allowance for an extra ⅜" at the front for the extra faceplate.

Next make the speaker cutouts according to Fig. 8-2. Table 8-1 lists materials. It's a good idea to make the cutouts in the front panel, the back one, and in one of the sides and then check the position of the mid-tweeter cutout in the other side. If you put each mid-tweeter cutout in the same position, one of the mid-tweeters will interfere with the installation of the super tweeter. The easy way to solve this problem is to mount the mid-tweeter in the side nearest the super tweeter in the middle of its panel. If you examine Fig. 8-6 carefully, you can see that this was done in the enclosure.

Set three of the panels together and measure the exact size needed for the partition, probably about 8-⅝" × 9" (Fig. 8-3). Cut this part to fit and drill two ¼" holes through it for speaker cords. Draw lines ½" apart through the middle of each panel to indicate the position of the partition. Support the panels on a firm surface and start #4-penny finishing nails along the mid-point of these lines.

Start a line of nails at each vertical edge of the front and back panel, located ¼″ from the edges. Start a line of nails ¼″ from the bottom edge of each panel. Drill two 11/64″ holes through the top edge of each panel. These holes are spaced 2″ each side of center, or 4″ apart. Then glue and nail a side to the partition. Glue and nail the front to the partition and the side. Install the second side and the back. When the glue sets, caulk the inner joints at the corners and around the partition.

Cut out the faceplate from ⅜″ plywood. Make the speaker cutouts according to Fig. 8-2. Check these by setting the faceplate on the enclosure and setting the speakers in the holes. The faceplate should fit the speaker board fairly closely at the edges, and the speaker holes should line up precisely. Outline the position of the mid-tweeter on the plywood with a pencil. For a perfectly flush fit you should remove about 1/16″ of wood from the surface of the

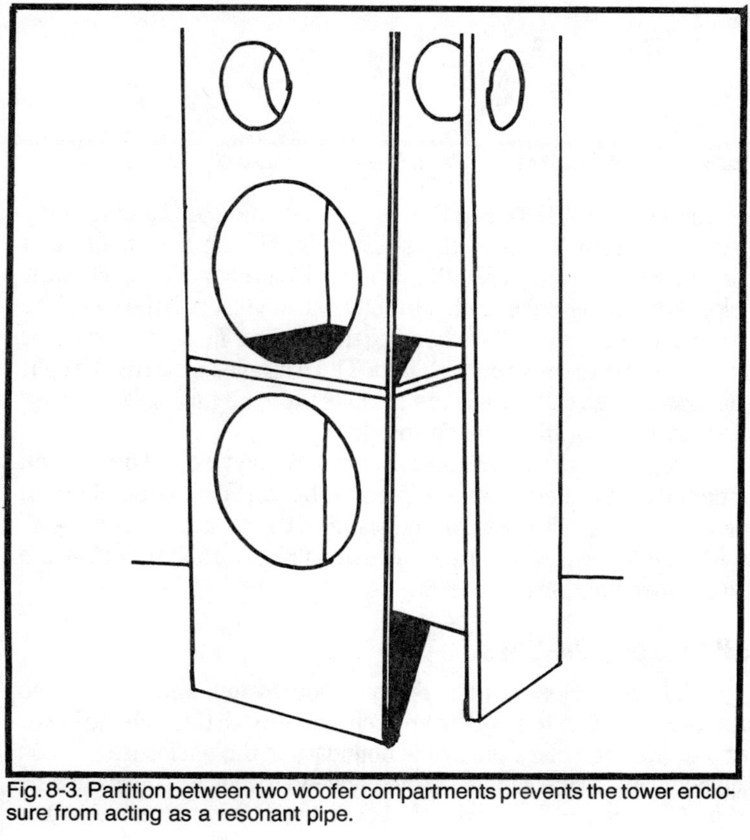

Fig. 8-3. Partition between two woofer compartments prevents the tower enclosure from acting as a resonant pipe.

Fig. 8-4. Check speakers for a flush fit on the enclosure.

plywood in a ring between the marked outline and the inner hole (Fig. 8-4). Test the fit of the speaker by placing a straight edge across the faceplate. This step is not absolutely necessary, but it helps to keep the grille cloth perfectly flat. Lightly nail the faceplate to the enclosure and, after checking the speaker fit again, remove it and apply glue to the speaker board. The nail holes made by the light nailing will precisely locate the faceplate for the final nailing. Paint the four sides of the shell flat black.

Prepare a 12″ × 12″ bottom from ¾″ plywood. Then cut an upper bottom piece from ½″ particle board. This upper bottom should theoretically measure the same as the partition, about 8-⅝″ × 9″, but there's a good chance you must remove about 1/16″ from each dimension for an easy fit.

ENCLOSURE GUIDELINES

Set the enclosure over the plywood bottom and center it so that the bottom extends 1″ from each side. Reach through the lower woofer hole and mark the inside boundary of the enclosure shell on the plywood. Make a mark on the plywood to indicate the front side of it. Glue and nail the particle board upper bottom piece to the plywood bottom, locating the particle board section so it fits just within the lines on the plywood.

Cut two pieces of lamp cord, one 24″ long and one 36″ long. Divide the conductors at each end and strip about ½″ of insulation from each conductor. Drill two ¼″ holes through the center of the particle board upper bottom piece, placed to match the distance between the lugs on a 2-lug screw terminal strip.

Pass the beaded lamp cord leads through one of the holes in the bottom piece and connect them to the lug on the terminal strip. Pass the two smooth leads through the other hole and connect them to the other lug. Solder the connections and install the terminal on the lower side of the particle board with glue and tacks. Then glue the bottom on with silicone rubber glue.

Prepare a top piece from ¾″ plywood. Set it on the top of the enclosure, centering it so that it extends 1″ from each side. Reach

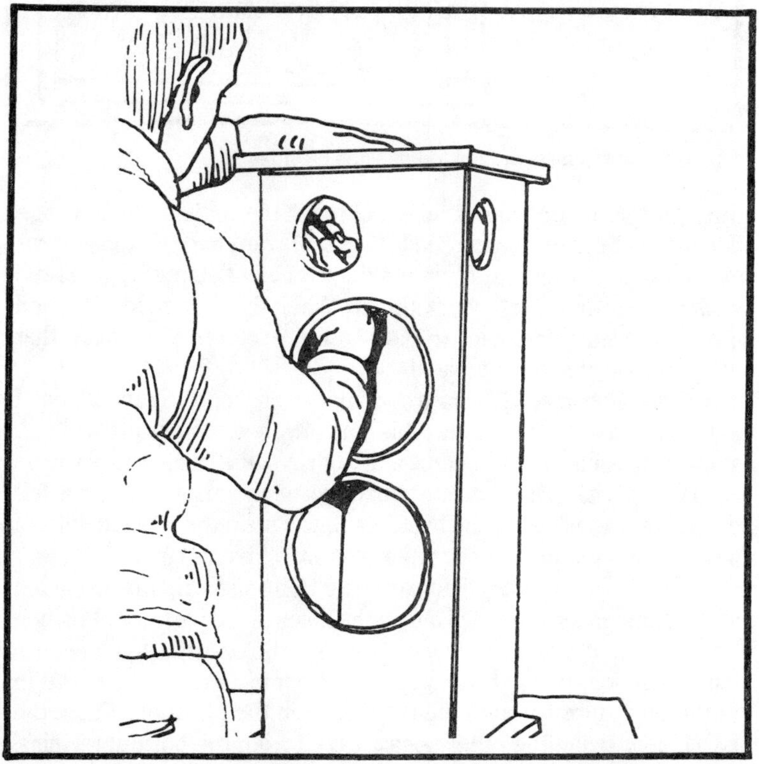

Fig. 8-5. Press the top down on the enclosure and mark the outline of sides under it to locate cleats.

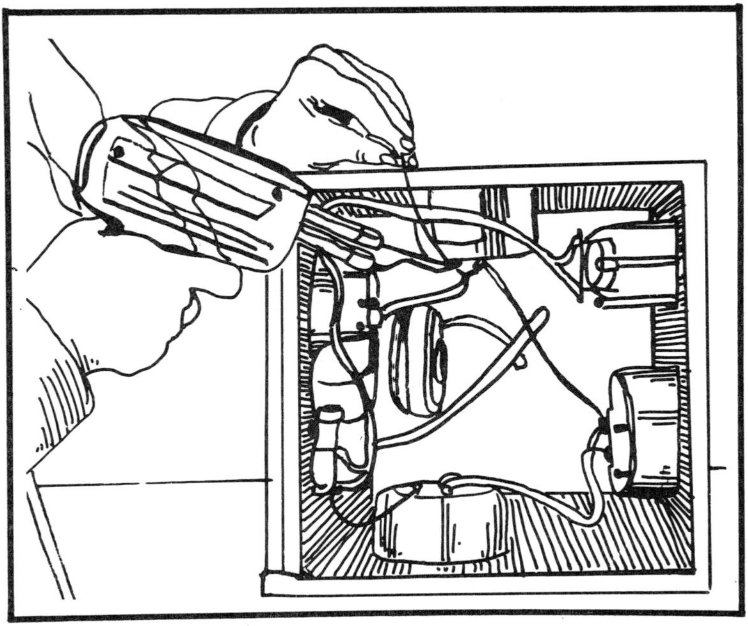

Fig. 8-6. Foam gasket around top of sides provides seal.

through the upper woofer hole and mark the outline of the inside boundary of the enclosure shell (Fig. 8-5). Remove the top and cut ¾″ × ¾″ pine cleats to fit within these lines. Temporarily nail the cleats into position and check the fit. When the fit is right, glue and install the cleats with nails (Fig. 8-7). Be careful to select nails that won't penetrate the top panel.

Glue a strip of ¾″ thin wood veneer around the raw edges of the plywood top and bottom. This kind of veneer comes in 8′ rolls. A single roll will do an enclosure, although in some cases there will be a small gap. Start the veneer at a back corner so that if any part is left uncovered, it will be at the back. Sand and stain the top and bottom pieces. Apply a finish coat of varnish or oil treatment.

Cut an extra piece of lamp cord 30″ long and strip the insulation at the ends. Pass it through one of the holes in the partition. Position this cord so it will reach the two woofer holes with equal excess length at each. Bring the longer piece of lamp cord from the bottom of the enclosure through the other hole in the partition. Caulk the two holes. Bring the other speaker cord from the bottom terminal out through the lower woofer hole. You should now have two speaker cords emerging from the lower woofer hole, one from the upper woofer hole, and an extra cord in the upper compartment (Fig. 8-6).

Loosely fill the lower compartment with polyester fill damping material. Set the enclosure on its back. Place one of the woofers on the faceplate and connect the two leads from the lower woofer hole to the lugs on that woofer. Carefully check to see that the two beaded leads go to the positive terminal on the woofer and the smooth leads to the negative terminal. Solder the connections. Run a bead of silicone rubber around the edge of the woofer cutout in the speaker board and set the woofer down on the glue. Twist the woofer slightly to get an even spread of glue under the frame; then press it down hard. Install another woofer in the upper compartment in the same way, checking to see that the beaded lead of the lamp cord goes to the positive terminal. Note that this lead should be the same one that goes to the woofer in the lower compartment.

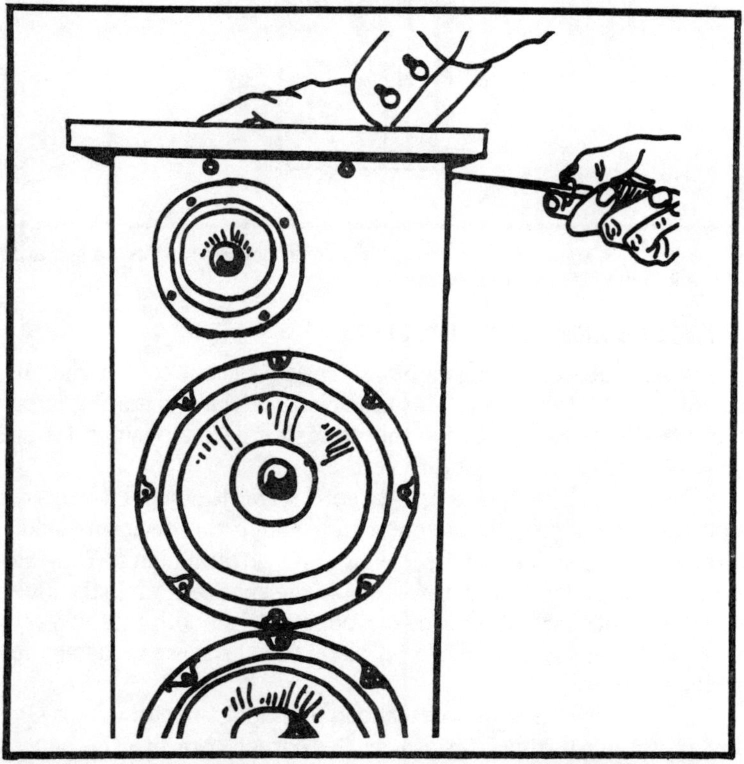

Fig. 8-7. Install top with screws driven through holes in sides into cleats under the top.

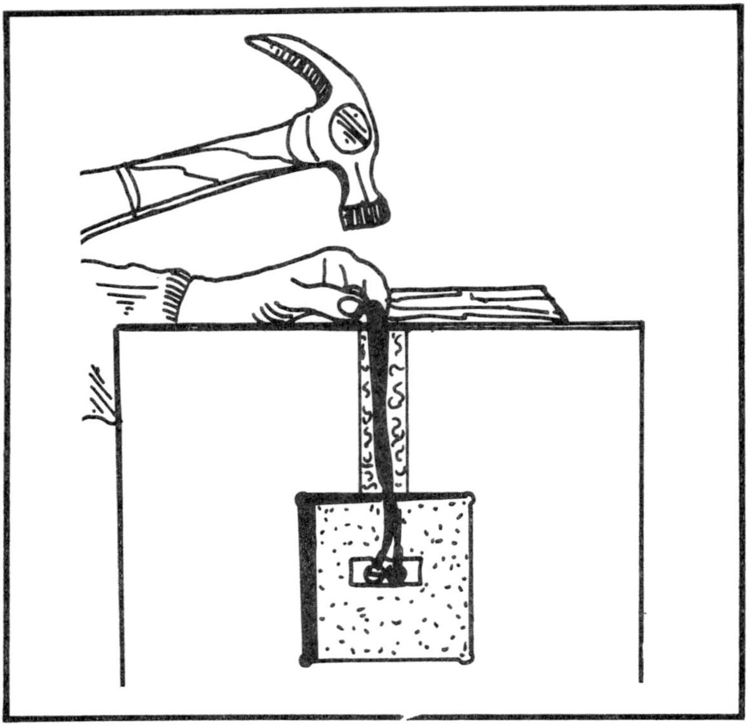

Fig. 8-8. Route speaker cable from terminals on bottom through the slot and staple it to the bottom's rear edge.

INSTALLING AND WIRING TWEETERS

Install the front facing mid-tweeter with silicone rubber glue. It isn't necessary to wire it at this time because you can reach it later through the open top. Leave the enclosure in this position for at least a half day for the glue to set.

Next install the other mid-tweeters and the super tweeter. You can mount the mid-tweeters by rotating the enclosure and, after gluing in each speaker, leaving it for several hours. You can hurry the work by using screws to hold the speakers while the glue sets. The silicone rubber doesn't adhere to smooth plastic very well, so it's a good idea to use ½″ sheet metal screws to install the super tweeter.

Next, wire the mid-tweeters and the super tweeter (Fig. 8-9). Install a 4-lug terminal strip near the corner opposite the super tweeter. Divide the mid-tweeters into pairs, if you can imagine the two on either side of the terminal strip as a pair. Connect the two

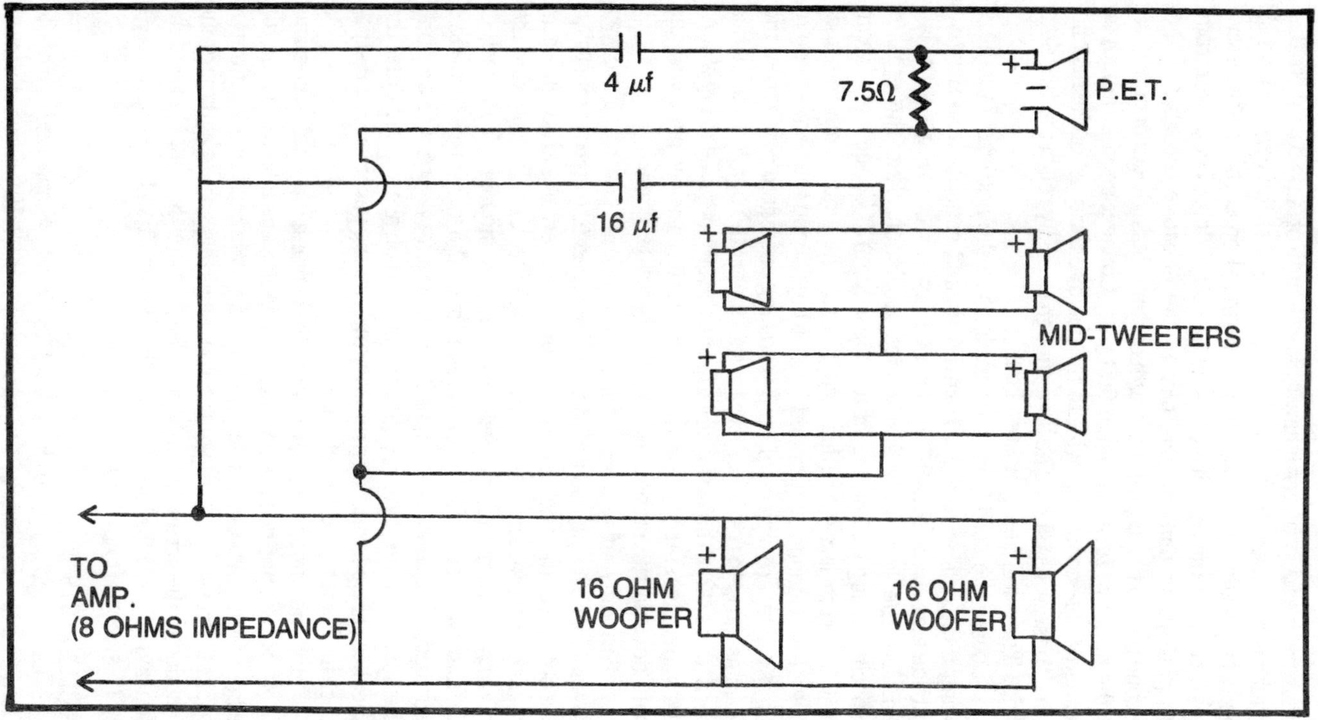

Fig. 8-9. Speaker wiring circuit for Project 18.

speakers in each pair by running a short length of lamp cord from the terminals of one to the other. Check to see that the beaded lead of the lamp cord goes to the positive terminal on each mid-tweeter. Next, connect the pairs in series by running a single conductor from the positive terminal of one pair to the negative terminal of the second pair. Then connect a length of smooth single conductor wire to the negative terminal of the first pair. Connect another length of beaded single conductor wire to the positive terminal of the second pair. These wires should be long enough to easily reach the terminal strip. They will be connected later.

Locate the cord from the terminals at the bottom of the box. Bring this cord up to the terminal strip and connect it to the two outside lugs. Connect a 16 μf and a 4 μf capacitor to lug #1, the lug that has the lead with the beaded insulation. Run the smaller capacitor to lug #2 and the larger one to lug #3. Connect a 7.5 ohm resistor from lug #2 to lug #4. Locate the leads from the mid-tweeters. Connect the smooth one to lug #4 and the beaded one to lug #3. Prepare a piece of lamp cord long enough to reach from the terminal strip to the super tweeter. Connect the beaded lead of this cord to lug #2 of the terminal strip and the smooth lead to lug #4. At the super tweeter connect the beaded lead to the terminal with the blue dot and the smooth one to the other terminal.

Turn off your receiver and turn the volume control all the way down. Connect a speaker cord from the receiver to the terminals on the bottom of the enclosure. Turn on the receiver and advance the volume just enough to hear the system easily. Check to see that the lows come from both woofers, the middles and highs from each of the mid-tweeters and the upper highs from the super tweeter. Give particular attention to the four individual mid-tweeters to see that the sound from each is at equal volume. If you are satisfied that the speakers are performing properly, turn off the receiver and disconnect it. Solder all connections.

Fill the upper compartment with damping material. Cut strips of ⅜″ × ¼″ open cell foam weather stripping and install it on the top edge of the enclosure sides so that the foam makes a continuous gasket. Press the top down firmly against this gasket and mark the location of the screw holes on the cleats under the top. Remove the top and drill small pilot holes for the screws. Install the top with #8 × 1″ screws (Fig. 8-7).

GRILLE CLOTH

Cut a piece of grille cloth about 28″ × 50″. Staple one end of the cloth to a back corner of the enclosure (Fig. 8-10). Wrap it around

Fig. 8-10. Staple grille cloth at a rear corner.

the enclosure and staple the other end over the first one. Go back and stretch the cloth vertically to reach the top and bottom of the enclosure.

The easiest and cheapest way to hide the raw edge of the grille cloth is to stretch a piece of ribbon around the enclosure at the top and bottom of the cloth. Or you can use pine quarter-round, as shown in Fig. 8-1. Each cabinet will require eight pieces of ½″ quarter-round, about 11″ long and cut at 45 degrees on the ends. The best way to get a good fit is to make the pieces a bit longer than 11″ and trim them to exact length. This method works best with a power saw. For hand mitering, it may be easier to try for a perfect fit the first cut.

To avoid staining the grille cloth, sand, stain and finish the trim pieces before installing them. Then attach the quarter-round with small brads to the upper surface of the bottom and the lower surface of the top.

When you move the completed speakers to your listening room, you should experiment a bit to find the best location for them. They are relatively tolerant of position, except when pushed tightly against a wall. The output of the super tweeter can be blocked there. You will find that you can alter the tonal balance of the system slightly by twisting the cabinets. When properly installed, the sound is amazingly uncolored even though these speakers are made with inexpensive components.

PROJECT 19: THE OMNI-HEX

- **Construction**: Most challenging.
- **Wiring**: Homemade filter.
- **Comments**: Fine cabinetry with smooth, well balanced omni sound.

Most omnidirectional speakers fire out at 90 degrees to each other from rectangular cabinets. Here is one that covers 360 degrees with better uniformity than would be possible if the same kind of speakers were put into a four-sided box (Fig. 8-11). Two of these cabinets will make handsome lamp tables in almost any living room.

The fine cabinet work will challenge even experienced woodworkers. But, as will be explained, a few changes in materials and procedure can simplify the cabinet work (Fig. 8-12 and Table 8-2). The simplified version will sound just as good. Only the appearance will be slightly less professional. And the cabinetry cost is somewhat lower in the second system.

CHOICE OF SPEAKERS

The cubic volume of one of these enclosures is about 3 cubic feet. Most modern high compliance 8″ woofers perform well in a half cubic foot closed box, so an obvious choice would be to put six such woofers and six small tweeters in each cabinet. The cost of such an arrangement provoked some second thoughts. Then it became apparent that the cabinet design would probably have to be altered slightly to accept front-mounted 8″ woofers. A change of about ½″ in the width of the trim on either side of the speaker boards would make space for such woofers, but the appearance of the cabinet would be changed. The woofers could be installed behind the speaker board, but such mounting isn't recommended for small tweeters. Moving the tweeters to the front of the board would accentuate the distance between speaker voice coils and increase phase shift. While these are not insurmountable problems, it seemed worthwhile to seek a simpler arrangement.

Fig. 8-11. The Omni-Hex adds a touch of elegance to any room.

Oval Speakers

After considering various alternatives, it became apparent that six oval speakers would solve the problem. A 6″ × 9″ speaker has approximately the same cone area as an 8″ speaker but, by virtue of its narrower dimension, has better dispersion. And if the oval speaker has a whizzer cone, it will give extended high frequency performance. Such speakers have been used for car radios for many years and are recognized as having several advantages over round

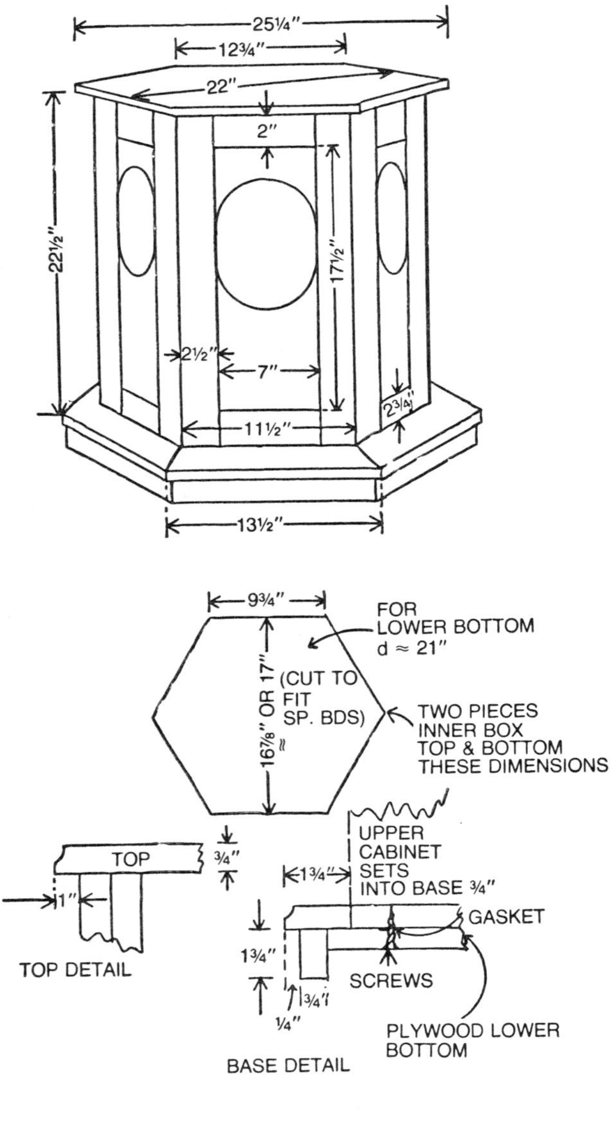

Fig. 8-12. Construction plans for Project 19.

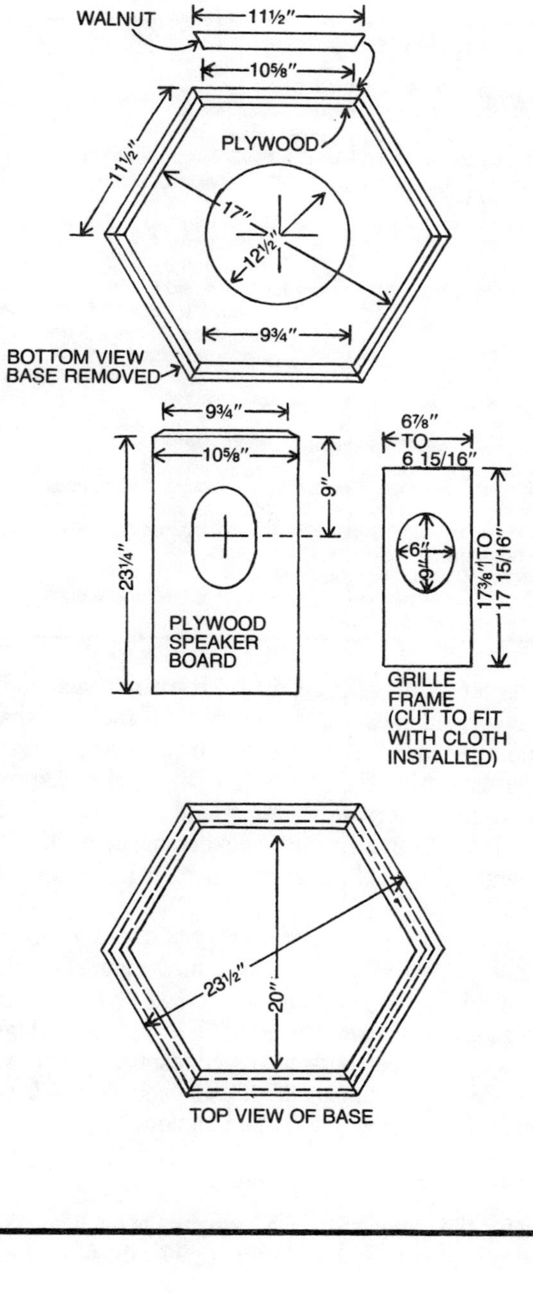

WALNUT
11½″
10⅝″
11½″
PLYWOOD
17″
12½″
9¾″
BOTTOM VIEW
BASE REMOVED
9¾″
10⅝″
9″
23¼″
PLYWOOD
SPEAKER
BOARD
6⅞″
TO
6 15/16″
6″
9″
17⅞″ TO
17 15/16″
GRILLE
FRAME
(CUT TO FIT
WITH CLOTH
INSTALLED)
23½″
20″
TOP VIEW OF BASE

Table 8-2. Materials list for Project 19.

Enclosure Materials		
¾" Plywood or particle board, pieces needed:		
2	top and bottom for inner box, octagonal pieces, about 17" across, 9-¾" per side.	
1	base oct. piece, about 21-½" across, 12-½" per side.	
6	10-⅝" × 23-¼", edges cut at 30°, speaker boards	
¾" Walnut or other hardwood.		
6	2" × 7"	Upper trim
12	2½" × 23¼"	Sides, trim
6	3-½" × 7"	Lower trim
1	Oct. piece, about 22" across, 12-¾" per side. Top.	
12	1-¾" × 13-½"	Lip around base and foot under base. Foot pieces will be shortened slightly.
½" plywood		
6	10-½" × 23-⅛"	grille frames.
Speakers and Components Needed		
6	6" × 9" full range speakers with magnet of 10 ounces, or greater, and whizzer cones.	
1	SPDT toggle switch, Radio Shack 275-402 or equivalent.	
1	6μf non-polarized electrolytic capacitor	
1	10 ohm, 10 watt, or greater, resistor.	
½ lb.	#20 gauge magnet wire. A ½" pound roll will make coils for two speaker systems.	

speakers. First, as noted, they can be installed on a narrower baffle. Second, when low-priced speakers are compared, the oval shape usually offers smoother sound. And when the oval speaker is installed with the longer dimension in a vertical line, the oval speaker will have better horizontal dispersion than a round speaker with the same cone area. Obviously good round speakers can be built. When price is a consideration, though, the oval speaker has the advantage.

Listening tests confirmed the advisability of choosing six oval speakers. The speakers selected had 10 ounce magnets, which should be considered a minimum. The cones had fairly compliant accordian suspensions. Somewhat greater bass range, at higher cost, could probably have been obtained by choosing speakers with foam suspensions. Most medium quality 6" × 9" speakers will work well in this project if they have adequate magnets.

Performance Advantage of Multiple Speakers

One reason for the satisfactory performance of such speakers is that when harnessed up with five others, or 11 others in stereo, each speaker carries a small fraction of the load. This means low distortion, much lower than it would be if a single speaker of the same kind were used for each channel. Some multiple speakers give erratic performance because the sound from adjacent speakers on a flat baffle can mix in a way that produces interference and reinforcement. This produces a peaky response curve. In a hexagonal system that possibility is reduced by the angle between adjacent speakers, by distance, and the cabinet barrier between other speakers in the system.

This system was widely approved by the listeners who heard it, more for its smooth response than its frequency range. Many audio authorities agree that the shape of the response curve is more important than the extent of its width. To shape the curve of this speaker, a special filter can be switched into the circuit. The filter, which gives the effect of smoother response on most program material, depresses the mid-range slightly. Its action is broad enough to work satisfactorily with almost any set of 6″ × 9″ speakers. Other than a smooth response, low distortion is probably another reason for listener approval of this system. Massive cone area insures lower distortion, and a stereo pair of these hex speakers has more cone area than that supplied by four 12″ speakers.

If you want to add tweeters, you should invert the speaker boards when you install them to make room in the upper boards for the tweeters. An easy-to-wire system would be to select six full-range speakers with foam suspensions and add six tweeters with a single capacitor high pass filter between the woofers and the tweeters. The tweeters should be wired in parallel-series with each other, the same kind of circuit as the six 6″ × 9″ speakers. This makes the net impedance of the tweeters 5.33 ohms, so the value of capacitance needed for a high pass filter for 8-ohm speakers should be increased by a factor of 1.5 to maintain the same crossover frequency. If you build such a system, get 6″ × 9″ speakers with the lowest possible frequency of resonance.

CHOICE OF MATERIALS

The enclosure pictured here was built from a combination of good quality ¾″ birch plywood and solid walnut (Fig. 8-13). The inner box, consisting of all the parts that don't show, was made of plywood (Fig. 8-14). All of the trim, the framing around the speaker boards, and the top were cut and shaped from pieces of solid walnut. If you use solid lumber for the exteriors, you must have facilities for joining boards together for the top. One possible alternative to this

Fig 8-13. The enclosure is ready for an oiled finish on walnut.

plan is to buy enough flooring boards to make the top. Flooring is available in oak and other woods. It will have ready made tongue and groove edges that you can fit together, but the pattern of individual boards will be noticeable.

If you don't want to go to the expense or the complexity of fitting together solid lumber, you can use good quality plywood throughout the project and cover the plywood edges with thin strips of wood veneer. Such a cabinet would lack the tooled edges shown here, but in other respects the appearance should be similar.

You can substitute ¾″ particle board for plywood in the construction of the inner box with no loss of quality. The plywood was used here because it was available and easier to rout for flush mounting of speakers. You can front mount the speakers, but that will require a narrower strip on each side of the speaker frames to support the grille cloth.

As you can see, the final choice of material depends on the tools you have available. If you have access to a router and joiner, you can duplicate the solid hardwood cabinet. If not, you must be satisfied with the plain edges of plywood.

CABINET CONSTRUCTION

You will probably have to juggle the dimensions of the individual parts of this project because of the many angles involved. If variations in dimensions are minor, you can adjust the cut of the trim pieces to make them fit together.

Start by cutting out parts for the inner box. When cutting the top or bottom pieces, either those of the internal or external pieces, you should cut opposite sides in sequence. For example, when cutting the inner top you should cut the width of the top, 16-⅞", first. If you cut one side, then progress around the piece by cutting 60 degree angles, you will compound any small error in your saw setting and end up far off the mark.

Next, cut the speaker boards to fit around the top and bottom. Each edge should be cut at 30 degrees so that two edges placed together make an angle of 60 degrees. For added strength, it's a

Fig. 8-14. Parts of the inner box for the Omni-Hex with drivers used in the project.

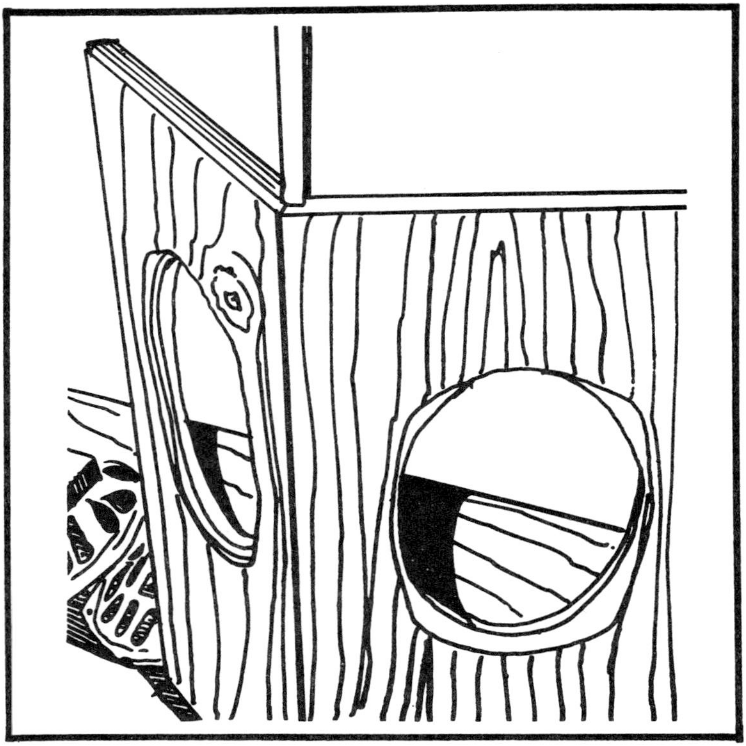

Fig. 8-15. For added strength, use splined joints on the inner box.

good idea to either spline the corners as shown in Fig. 8-10 or use 60 degree glue blocks behind each corner. You can make splines from any strong solid wood (Fig. 8-15). The dimensions of the splines should be about 1/8″ by about 3/4″. Cut a 1/8″ × 3/8″ groove in both edges of each speaker board to receive the splines. To cut the grooves use a radial arm saw or a table saw. Don't use a dado blade for these cuts; it might chip out the plywood edges. Use a saw blade and move the stock far enough aside to make the width of the cut right. Set up the boards and check the fit of the splines. If you get the grooves too large, you can make thicker splines. You may find that humidity changes will affect the fit of the splines.

Cutting Holes

Cut the speaker holes before assembling the box. For flush mounted speakers you will have to make two templates, one for the oval speaker hole and a second one to guide your router (Fig. 8-16).

The oval pattern can be made by a cut and try method until you get the hole in the pattern. Cut these holes with a router or a saber saw. For the other template, simply place the speaker face down on the template material and draw the outline of the frame on the template. Choose a smooth material for the router template. Formica makes a good template if it is backed by a stiff material, such as ⅛″ hardboard, or any suitable material that is no more than ¼″ thick.

Cut an access hole in the bottom panel about 12-½″ in diameter. Make a final check to see that all parts fit; then start gluing and assembling the inner box. You can nail the speaker boards to the inner top and bottom, but the splines and glue are adequate to hold the corners firmly together. You may have to drive the splines into the pre-glued grooves, particularly if the humidity is high when you assemble the box. Apply clamps in the speaker cutouts to maintain pressure on the joints while the glue sets.

Fig. 8-16. Make up a template, like the one on the left, to guide your router in removing enough wood from the speaker panel for flush mounting.

Making a Frame

Run the boards for the top through a joiner and glue them together under pressure. Again, when cutting the top to size, you should cut the width of the opposing sides first. In this case cut the piece to a width of 22″; then move to the next set of opposite sides and so on. When the top has been cut to size, you can put the ornamental groove in the upper corner with a shaper or router. Rout the end grain sides first. You will probably rough the end grain or chip it at the beginning or end of routing that side. Your final routing with the length of the grain will eliminate those rough edges. Glue and screw the top to the inner box with screws from inside the box.

Next, cut the trim pieces to fit around the edges of the speaker boards. These pieces should be 2-½″ wide for the vertical strips, 3-½″ wide for the bottom horizontal pieces, and 2″ wide for the top ones. Although the bottom pieces are 3-½″ wide, only 2-¾″ of them will show in the finished cabinet because the cabinet sets down ¾″ into the lower base.

After the pieces are cut out, glue them together to make a frame. This frame is much like a picture frame except that the corners have no beveled cuts. Make the angled cuts at the edge of each frame so the frames fit together around the inner box. Number each frame and each corresponding side because you will surely have to make minor adjustments to each frame to fit. When all pieces fit, attach the frames to the inner box with glue and screws driven from inside the box into the frames.

Cabinet Assembly

Next, cut out the lower bottom. Again, follow the opposing sides procedure. Cut the 21-½″ dimension first and then the other sides in pairs. Set the upper box on the lower bottom and mark its outline on the base. Next, prepare enough stock for the upper "lip" of the base by running it through your shaper—or by using a router to groove its top surface to match that of the top. Temporarily install the base on the cabinet with screws; then cut the lip material to fit around the cabinet. When the pieces fit snugly, glue the lips on the base. Use care to prevent glue from getting between these trim parts and the upper cabinet. When clamped, you can separate the base from the cabinet to insure that the two parts are not glued together.

Next cut the feet to fit under the base. Attach them to the base with glue, again clamping them until the glue sets. This completes the assembly of the cabinet except for the speaker grille frames and installing the speakers and the base. You can now finish sanding the exterior parts and apply stain and the final finish to them.

Make grille frames to the dimensions shown in Fig. 8-12 and install grille cloth on them. Choose an open weave cloth. It's a good idea to use a piece of carpet or old blanket on your work bench to protect the finish of the cabinet. Install the speakers in the box, one at a time, with silicone rubber caulking compound. Position the cabinet with the board to receive each speaker facing upward; then run a heavy bead of silicone rubber around the speaker opening. Set the speaker on the board and jiggle it slightly to press the silicone rubber into every space between the speaker frame and the board. Leave the cabinet in that position until the silicone rubber sets up. If you want to hurry the work, you can use screws to hold the speakers against the silicone rubber glue and install the speakers at once.

WIRING SPEAKERS

When you wire the speakers, an inside light is a great help. Put the box on its side with a protective surface under it (Fig. 8-17).

Fig. 8-17. Set upper box on a rug or other padded surface to protect finish while wiring speakers.

Insert a socket with a small wattage bare bulb into the box. You can hang the socket by attaching it to the upper side of the magnet of one of the upper speakers. It will provide the light you need to avoid mistakes while wiring the lower three speakers in parallel. Then roll the cabinet until the other three speakers are on the bottom, and wire them in parallel.

Connect a wire from the plus terminal of the first group to the minus terminal of the second group of three (Fig. 8-18). Join a 24″ lead to the minus terminal of the first group. Connect another 24″ lead to the plus terminal of the second group. Check for proper polarity by connecting a 1.5-volt battery to the two leads and watching the speaker cones.

All six cones should move out and together when the connection is made and broken. Solder all connections with rosin core solder.

MOUNTING FILTER COMPONENTS

Because of the available space under the base, the filter components were installed there. If you prefer to place them inside the cabinet, you can make a cutout in the base and install the speaker terminals and filter components on a piece of ¼″ tempered masonite. This was done with crossover networks in many of the projects in this book. External mounting permits changing components and, more important, it puts the switch near one corner of the base where it can be easily reached. If you use the cutout method, the cutout should be located some distance away from the edge of the base. In that case you would probably have to lay the cabinet on its side to use the switch.

To mount the components under the base you will need a three-screw terminal strip for the speakers or a couple of two-terminal strips (Fig. 8-19). You need one extra terminal screw to receive one amplifier lead. You can wire the filter between it and one of the two speaker terminals. Drill a ¼″ hole through the base for each screw terminal. Feed the ends of the 24″ leads through two of the holes and solder them to the terminals on the outside of the base. Install the terminal strip(s) with silicone rubber cement and small screws or tacks. Check to see if the speakers are working when you connect a signal source to the terminal strips. Avoid driving the speakers hard with the base off the box.

Fill the box with loose polyester fiber. Some kinds of batting must be stapled to the top and side walls of the inner box to prevent settling. About 2 to 3 pounds of material per cabinet will do the job.

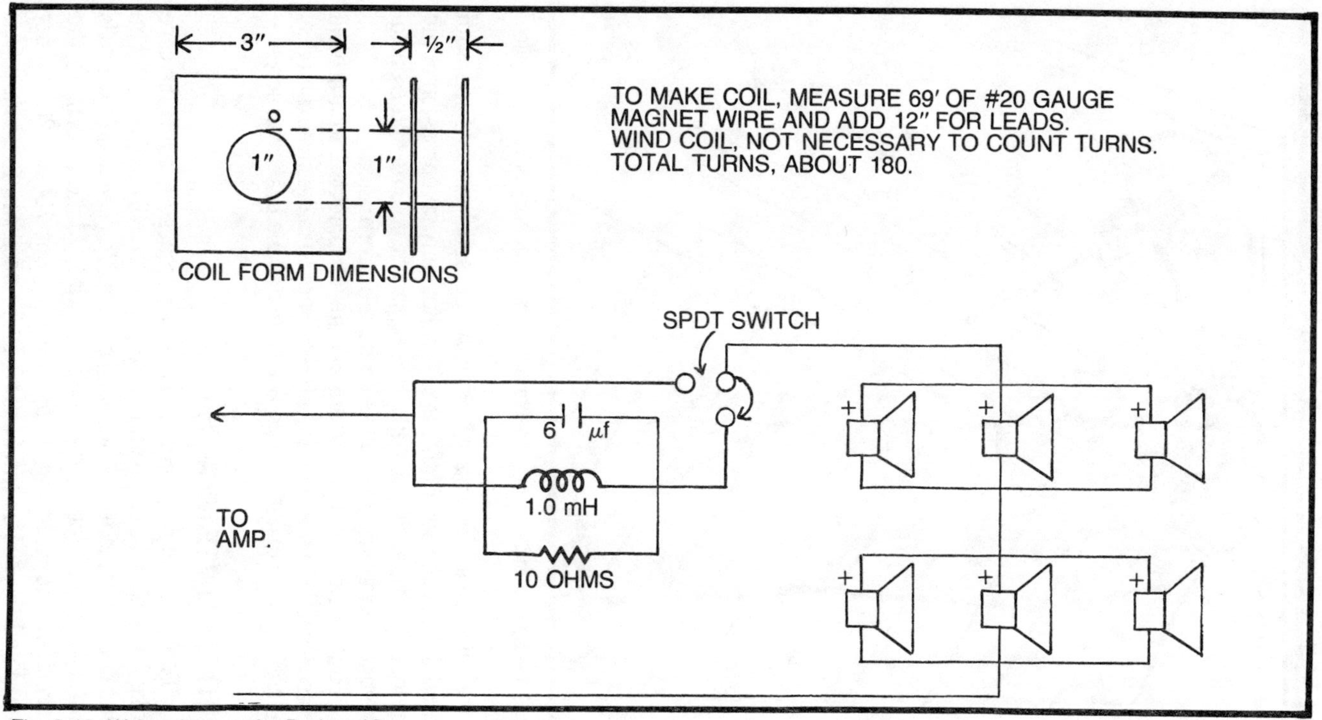

Fig. 8-18. Wiring diagram for Project 19.

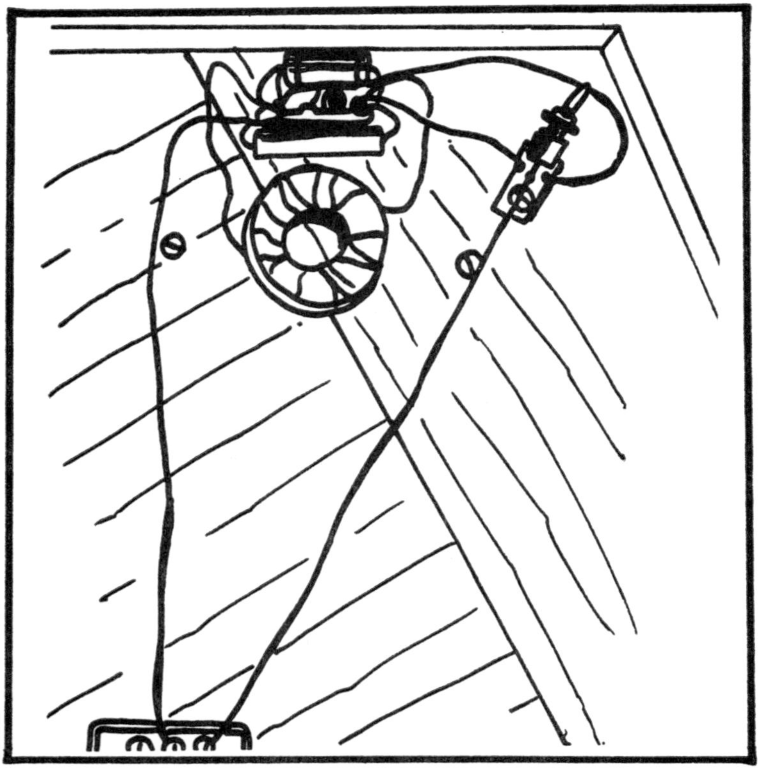

Fig. 8-19. Layout of filter components under base for easy access to switch.

Make sure that the space between the speakers is filled with material. Fill each box until it is full without any compression of the material. A proper fill will add to the apparent volume of the enclosure and permit slightly improved bass performance.

Run a strip of motite or foam weather stripping material around the access hole in the bottom of the cabinet to make an airtight seal when the base is installed (Fig. 8-20). Install the base with screws.

THE FILTER AND FILTER ACTION

To make the filter you can either purchase a 1 mH air core choke for each system you build, or you can easily make one. The value of this component is not at all critical. You can make two coils from a ½ pound roll of #20 magnet wire. Make a form to the dimensions shown in Fig. 8-18. Measure out 70′ of #20 magnet wire. Wind the coil on the form. It isn't necessary to count turns.

The final value will be close enough if you wind it with even a reasonable amount of care to see that no loose loops or crude air gaps are made. If you make the coil form by bolting together the parts, you can extract the coil from the form and tape it. Use the form to wind the coil for your next speaker.

Almost any small switch will do to cut the filter in and out of the speaker circuit. Here an SPDT toggle switch was used. The reason for the switch is that some program material may be deficient in the frequencies that are depressed by the filter. Some listeners may prefer to eliminate the filter most of the time, but those who heard this system voted for the filter with statements such as, "If it must be on or off all the time, leave it on." When assessing the action of the filter, be aware that it depresses the sound volume slightly. Your ear will usually tell you that the louder of two speakers is the better. So, to be fair, the volume control on the amplifier should be adjusted upward to compensate for the action of the filter when it is switched in. A good rule of thumb is to use the filter until you hear music where a vocalist is lost among the instruments. Then switch off the filter.

Experience with the speakers thus far indicates that the filter should be used about 90 percent of the time. This filter was used successfully with two kinds of 6″ × 9″ oval speakers. If you want to

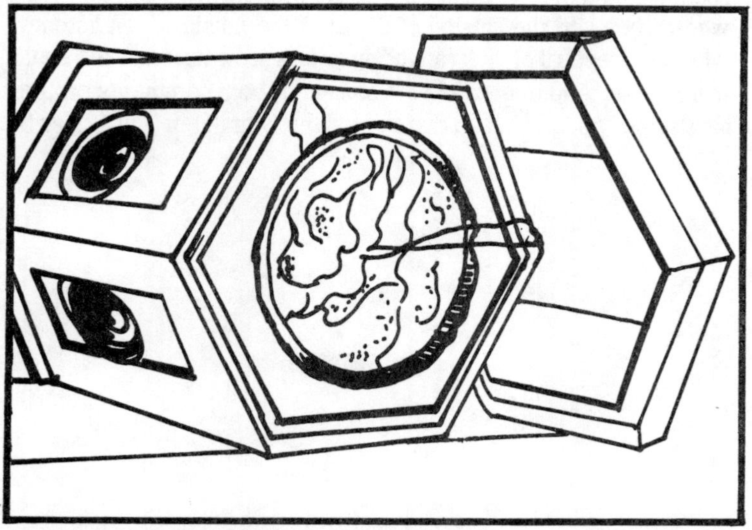

Fig. 8-20. Fill enclosure with loose polyester damping material. Run a bead of mortite, shown here, or foam weather stripping around the access hole before installing the base.

experiment with the filter, you can adjust the frequency by altering the value of the capacitor or its effectiveness or sharpness by changing the value of the resistor. A larger capacitor will lower the filter's frequency range; a smaller value will raise it.

If you make a significant shift in the frequency range, you will also change the value of the reactive components at the affected range. Here are some figures to guide you. The action of the filter shown in Fig. 8-18 is centered at about 2000 Hz, where the net reactance is about 13 ohms before adding the resistor. If you substitute a 12 μf capacitor, the frequency will drop to about 1450 Hz, and the reactance to about 9 ohms. If you decrease the value of the capacitor to 3 μf, the frequency most affected will be 2900 Hz. The reactance there will be about 18 ohms. In any case remember that the values shown in Fig. 8-18 were chosen by listening tests, not by calculations. Considering that, you may want to experiment with different values of resistors. The purpose of the resistor is to broaden that filter action and prevent a hole in the response curve. The best way to choose the right resistor for any situation is by listneing. Don't be afraid to experiment, but begin by using the filter specified in Fig. 8-18.

When the filter action satisfies you, the project is finished. You will find that these speakers look good and sound good at almost any room location. They can set out in the room, away from the walls, where they add the illusion of depth of the music. Even listeners who are skeptical of "full range" speakers are amazed at the sound from these. And everyone who has seen them comments on their handsome appearance. In that regard the Omni-Hex is in a class by itself.

Van and Car Speakers

A recent development in van and car speakers is the miniature woofer/tweeter system in a compact sealed enclosure. Such speakers have extended the range of high fidelity performance in vehicles. In fact, their quality is so high that many buyers use them for home music systems where space is limited. The drivers used in the two projects here meet those requirements perfectly. By building your own system you will have a set of speakers whose performance will equal that of commercial systems costing about twice as much.

The same drivers are used in the two projects. They can be built into a ported enclosure for maximum bass range or into a closed box that is slightly smaller. Norelco, the manufacturer of the drivers, suggests the closed box version for cars, pointing to the fact that the inner dimensions of cars are not great enough to permit the reproduction of frequencies below about 100 Hz. So here are the projects—a ported system with full crossover network as a top-of-the-line van speaker and a closed box speaker for cars. Either of these speakers will make a fine choice for a compact home music system of impeccable quality.

PROJECT 20: THE VANNER

- **Construction:** Use power saw. Can simplify.
- **Wiring:** Commercial crossover.
- **Comments:** Amazing. 4 ohms impedance.

Fig. 9-1. Van speaker with drivers.

In keeping with the level of performance offered by these speakers, the obvious choice of building material seems to be hardwood plywood (Fig. 9-1). You can use particle board and get just as good a sound. If you use butt joints, you must adjust the dimensions of the parts to give the same internal dimensions that are shown here.

The usual method of joining hardwood plywood panels is to make 45 degree cuts, but for the van speakers the *Porter joint* is used. One reason for this choice is even if van speakers are hung on a bracket, they are likely to take more punishment than house speakers. The Porter joint puts a strip of solid wood in each external corner of the box, making a corner that can withstand knocks without splintering. It is relatively easy to make, and the appearance is very good.

Some lumberyards will sell birch plywood only in full 4′ × 8′ sheets. If you can't find a dealer who sells partial sheets, you can try to buy scraps from cabinet shops or even from house builders. Note that the enclosure requires ¾″ plywood for all panels except the

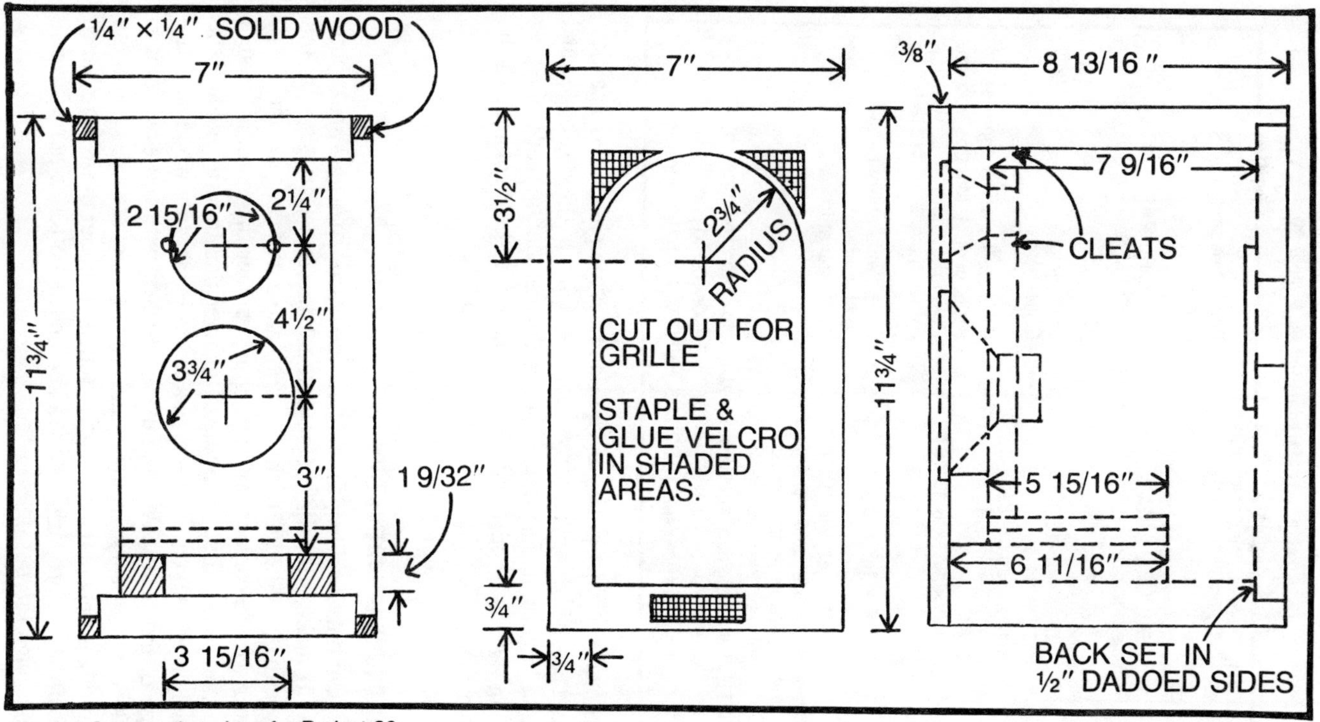

Fig. 9-2 Construction plans for Project 20.

Table 9-1. Materials list for Project 20.

¾″ Birch Plywood, pieces needed:		
2	8-13/16″ × 6-½″	Top & bottom
2	8-13/16″ × 11-¼″	Sides
¾″ Plywood or particle board:		
1	5-½″ × 9-21/32″	Speaker board
½″ Particle board.		
1	6-½″ × 11-¼″	Back,
⅜″ Plywood		
1	5-½″ × 5-15/16″	Duct panel
1	7″ × 11-¾″	Grille frame
Solid wood, pieces needed:		
2	13/16″ × 19/32″ × 6-11/16″	Glue blocks which serve as sides of duct. Must be accurate
4	¼″ × ¼″ × 8-13/16″	Exterior corners. Can substitute ¼″ quarter-round.
2	½″ × ¾″ × 9-½″	Side cleats at front, cut length to fit space.
1	½″ × ¾″ × 4-½″	Top cleat at front.
Damping Material, acoustical fiber glass; 60 grams fiber glass. About 1″ × 12″ × 18″		
NOTE: If alternate construction is used, inside dimensions should be 5-½″ wide × 10-¼″ high × 7-9/16″ deep and duct should be precisely 19/32″ × 3-15/16″ × 6-11/16″ long.		

back and the duct panel. If you don't have a piece of ½″ particle board for the back, you can extend the depth of the box by ¼″ and use ¾″ material. The duct panel can preferably be made of either ⅜″ or ½″ plywood for it. The reduction in box volume will be insignificant. See Fig. 9-2 and Table 9-1.

ENCLOSURE CONSTRUCTION

Cut out the sides, the top and the bottom pieces. Adjust your saw to make a cut ½″ deep, and set the fence at ½″ from the opposite cutting edge of the blade. Make a trial cut on some waste material to check your saw setting. By running the material through the saw twice, first in the usual manner and then by holding the panel on edge, you will remove a ½″ × ½″ strip of wood from the edge of the panel. When you are satisfied with the saw setting, make cuts at the upper, lower and rear edges of the sides and at the rear edges only of the top and bottom pieces. Set these together to see if they fit properly. Mark a line on the bottom where the side panel joins the bottom.

Next, cut the glue blocks that fit into the lower joints and position the duct panel. These blocks must be made with precision because their dimensions will determine the exact cross-sectional area of the duct. The dimensions listed in the plans are 13/16″ wide × 19/32″ high. The figure for the height of the glue blocks, 19/32″ or 15 mm, will be the correct measurement regardless of any minor variation in your box dimensions. You may have to alter the width of the glue blocks to maintain the right duct width which is 3-15/16″. When you have cut the glue blocks to the proper dimensions, glue and screw or nail them to the bottom, using the previously made lines as guides to locate the blocks.

Cut a duct panel and install it over the glue blocks using glue and nails (Fig. 9-3). Notice that the duct panel is recessed ¾″ from the front edge of the glue blocks to make room for the speaker board. Check the fit of all parts at each step to make sure no part will fit too loosely or protrude and displace other parts. You can fill minor gaps with caulking material.

Set the sides on the bottom and mark the sides where the top of the duct panel joints them. Mark a line, indented ¾″ from the front edge, on the inner wall at each side. Cut two side cleats to fit in the space between the duct panel and the top. Glue and screw or nail these cleats to the sides, placing them behind the lines previously made.

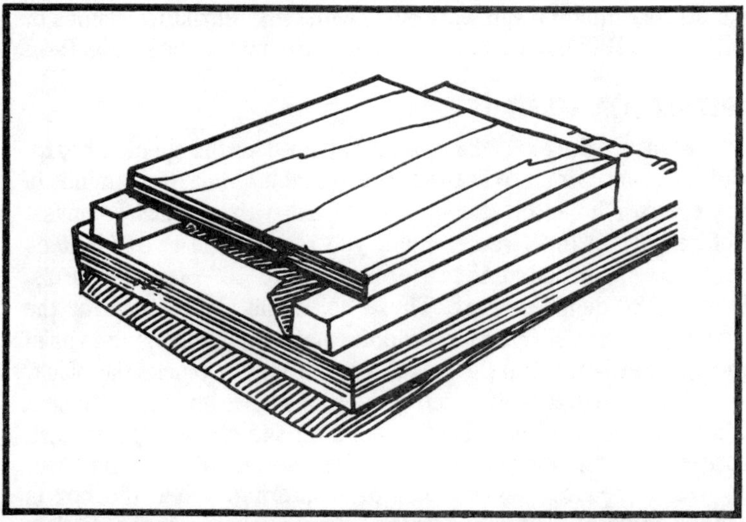

Fig. 9-3. Assemble bottom. Glue blocks and duct panel first. Duct dimensions must be those shown in Fig. 9-2.

Fig. 9-4. A couple of old 20″ bicycle tubes are just right for holding enclosure sides under pressure while gluing.

Apply glue to each joining surface of the enclosure. Use enough glue to cover each surface when spread thin; too much will seep out of the joints onto the unfinished plywood and stain it. Assemble the panels together and apply pressure with either furniture clamps or with two old 20″ bicycle tubes stretched around the shell. (Fig. 9-4).

SPEAKER BOARD DETAILS

Measure the exact dimensions required for the speaker board. Cut it out and trim it to fit the space available. Mark the outlines of the speaker holes on the speaker board with a pencil compass. Before cutting the tweeter hole, you should drill two 3/8″ holes centered on the circumference of the 2-15/16″ diameter circle, located 180 degrees apart. These holes will make room for the tweeter terminals. Cut out the woofer and tweeter holes; then paint the speaker board and the front edge of the glue blocks flat black.

After the glue sets, remove the tubes again and sand the box. It's a good idea to just slightly round off the sharp edge at each corner. Such rounding makes the cabinet feel smoother to the touch as well as reduces the chances of splintering. When the box is sanded smooth, stain the exterior to any color you desire. Follow the stain with an oil finish or varnish (Fig. 9-5).

Install the speaker board. Run a ¼″ bead of silicone rubber glue around the cleats and on the front edge of the duct panel. Press the board down firmly onto the sealer. If you don't have silicone rubber sealer, you can use butyl or latex caulking material.

SPEAKER ATTACHMENT

Install leads on the speakers before you mount them. The holes in the terminals on the woofer are unduly small. If you are using lamp cord, you can wrap a bared section of wire around each terminal, crimp it with long nose pliers and then solder the connection. Pay attention to which conductor goes to the positive terminal of each speaker. Notice that the red dot terminal is on the left for the woofer and on the right for the tweeter. Use the ribbed lead of our lamp cord as a guide to the positive terminals on each driver.

Install the speakers by running a bead of silicone rubber sealant around each cutout and pressing the driver down on the caulking material. Leave the box on its back until the rubber has set firmly, a half day or longer.

Cut out the back to fit the space available for it approximately 6-½″ × 11-¼″. Notch the corners of the back so it will fit over the inside corners of the solid wood strips at each joint. Make a 2″

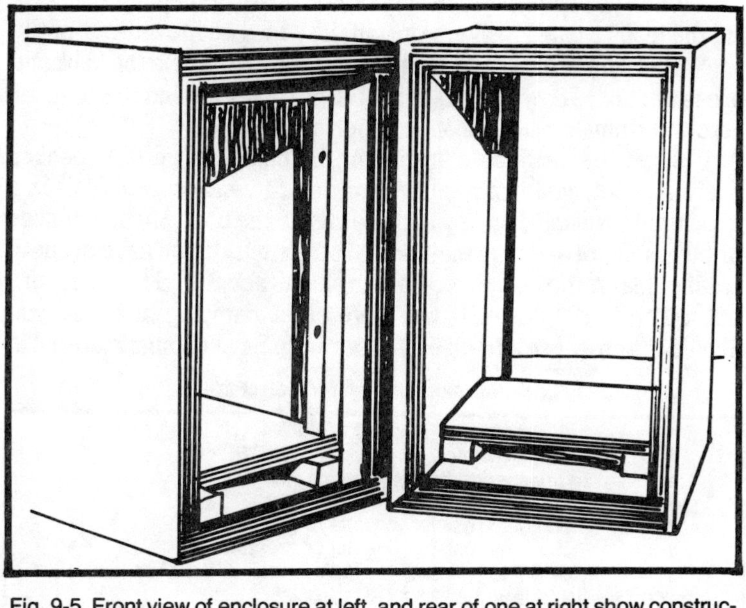

Fig. 9-5. Front view of enclosure at left, and rear of one at right show construction details.

horizontal × 3″ vertical cutout in the back, centered behind the woofer, to give access to the crossover board.

CROSSOVER BOARD WIRING

Prepare the crossover board according to Fig. 9-6 and install the crossover network on it. You will probably have to remove short projections on the bottom of the board with your side cutting pliers before gluing down the board with a blob of silicone rubber at each end. Install a terminal strip in the position shown by gluing it down with heavy adhesive. Mount the screw terminal strip on the exterior of the board with glue and tacks. Install the toggle switch in the preapred ½″ hole. Wire the board according to Fig. 9-6. See Table 9-2 for components.

The two resistors, a 3.5-ohm series resistor and a 7.5-ohm shunt resistor, permit you to attenuate the tweeter's output by a pre-determined degree while maintaining the proper impedance in the tweeter circuit. A somewhat similar effect could be obtained by use of a tone control, but not precisely the same. A tone control would roll off the upper highs more than the middle highs, but the control on the speaker reduces the level of the tweeter's output equally at all frequencies carried by the tweeter.

When you have wired the crossover board, except for connecting the leads to the speakers, install the board on the interior of the back. Run a bead of silicone rubber around the cutout in the back and press the board down on the back. Fill the holes around the speaker screw terminals with silicone rubber.

When the caulking material on the back and on the speaker board has set, you can resume work on the enclosures. This is a good time to install damping material inside the box. About 2 ounces or 60 grams, of damping material is about right. If you have acoustical fiber glass, the kind that comes in sheets about 1″ thick, measure out a 12″ × 18″ piece. If you have other damping material, you should weigh it. Install it on each inner wall. Save enough material to

Table 9-2. Components list for Project 20.

1	4″ woofer, Norelco AD 4060/W4
1	Dome tweeter, Norelco AD0163/T8
1	Crossover network, Norelco ADF 3000/4
Optional control network:	
1	3.5 ohm resistor, 5 watts or greater.
1	7.5 ohm resistor, 5 watts or greater
1	SPDT toggle switch, Radio Shack 275-603 or equivalent.
Speakers available from Gladstone Electronics, 1736 Avenue Rd., Toronto, ON M5M 3Y7	

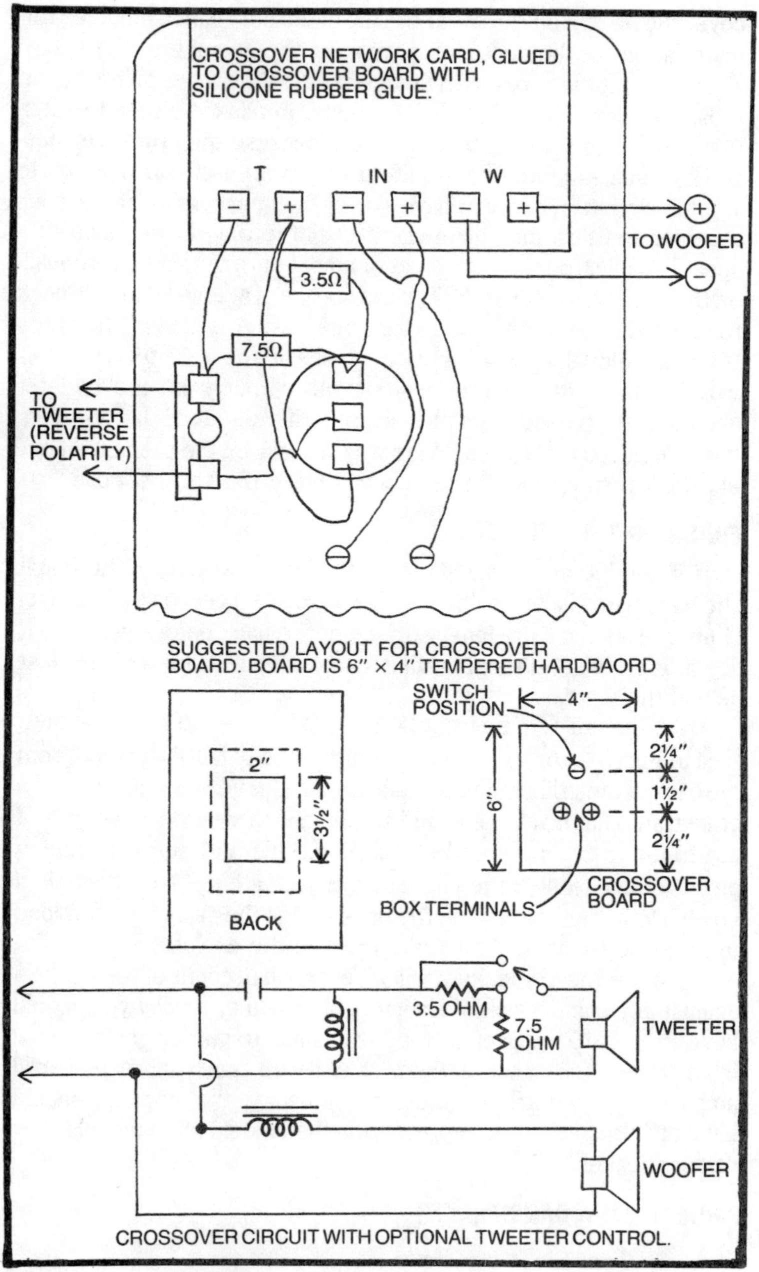

Fig. 9-6. Crossover wiring for Project 20.

cover the back. You can cover the top of the duct panel, but keep the material out of the duct and away from the opening to the duct.

Connect the wires from the speaker to the proper points on the crossover network and solder the joints. In this circuit the tweeter is wired out of phase with the woofer because the crossover network produces more phase shift in the tweeter circuit than in the woofer circuit. When you have wired the speakers, turn off your receiver and turn the volume control all the way down. Connect a speaker cable from your receiver or amplifier to the screw terminals on the exterior of the back. Try the speakers at low volume. Check to see if the lows are coming from the woofer and highs from the tweeter. Flip the toggle switch to see if the resistor circuit is working properly. At one position the output from the tweeter should be depressed. Don't be surprised if the sound is lacking deep bass when you test the speakers with no back on the cabinet. Above all, don't turn up the volume control while the back is loose.

FINISHING THE PROJECT

When you are satisfied that everything is working right, install the back with silicone rubber sealant. Don't forget to staple a layer of fiber glass over the inner surface of the back before you glue it down. If possible, it's a good idea to make the thickness double just behind the woofer.

Cut the grille frame from 3/8″ plywood. Note the cross-hatched area in Fig. 9-2. At those points remove about 1/16″ of wood from the back of the grille frame to make a space for velcro tabs. Paint the grille frame flat black. Glue and staple tabs of velcro to the back of the frame in the prepared slots. Also, glue and staple matching pieces on the front of the enclosure. Select a thin polyester cloth, a stretch knit, and staple it on the back side of the grille frame, folded over at the corners. Attach the grille to the enclosure.

Two of these speakers will give a good account of themselves against any competition of similar cubic volume. Ideally you should have at least 10 watts of power per channel to get full performance from them. You can use them safely with much more powerful amplifiers. If you are using them in a house, the amplifier should have an infrasonic filter, a precaution that should be taken with any reflex speaker.

PROJECT 21: A CAR SPEAKER

- **Construction:** Easy
- **Wiring:** Simple. Single capacitor high-pass filter.

Fig. 9-7. Car speaker with grille cloth installed.

- **Comments**: Similar to #20, but slightly reduced bass range. 4 ohms impedance.

This is a bonus project (Fig. 9-7). In comparing it with #20, and they are bound to be compared, one would have to say that this one has an amazing range for its size, while #20 is just plain amazing. If you are going to use the speaker in a car, this is the logical choice. For a van or a small room, #20 is worth the extra expense of the crossover network and the larger enclosure. Of course you can compromise and build a vented enclosure with the simple crossover of this project. Or you can use the crossover network with the closed box. The main advantage of the Norelco crossover network is that it gives better protection to the tweeter than the simple high-pass filter.

BUILDING TIPS

Here, in the last project in the book, is the same construction prodcedure that was used in the first one. The building material is ½″ particle board covered with an inexpensive plastic veneer (Fig. 9-8 and Table 9-3).

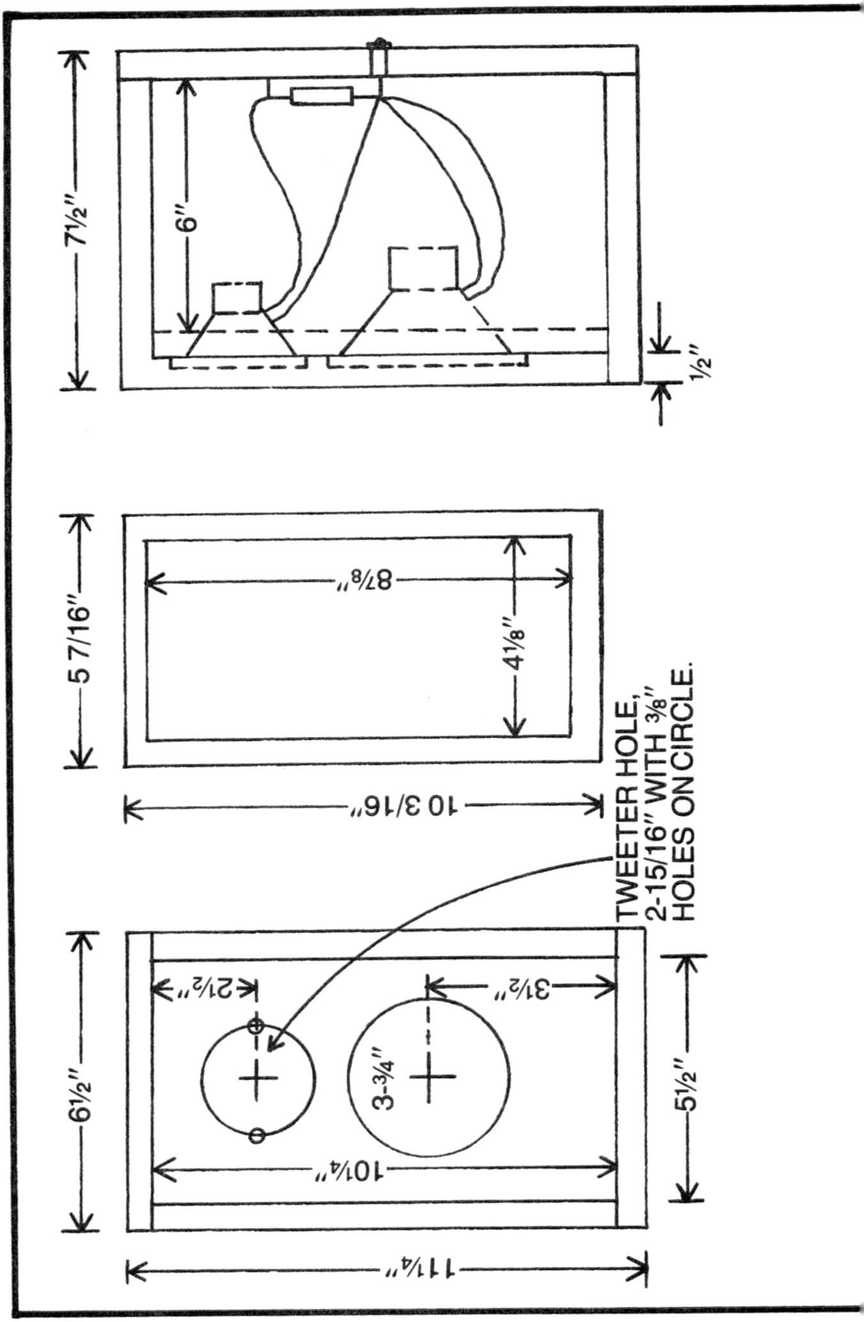
7½"
6"
½"
5 7/16"
8⅞"
4⅛"
10 3/16"
TWEETER HOLE,
2-15/16" WITH ⅜"
HOLES ON CIRCLE.
6½"
2½"
3½"
3-3/4"
5½"
10¼"
11¼"

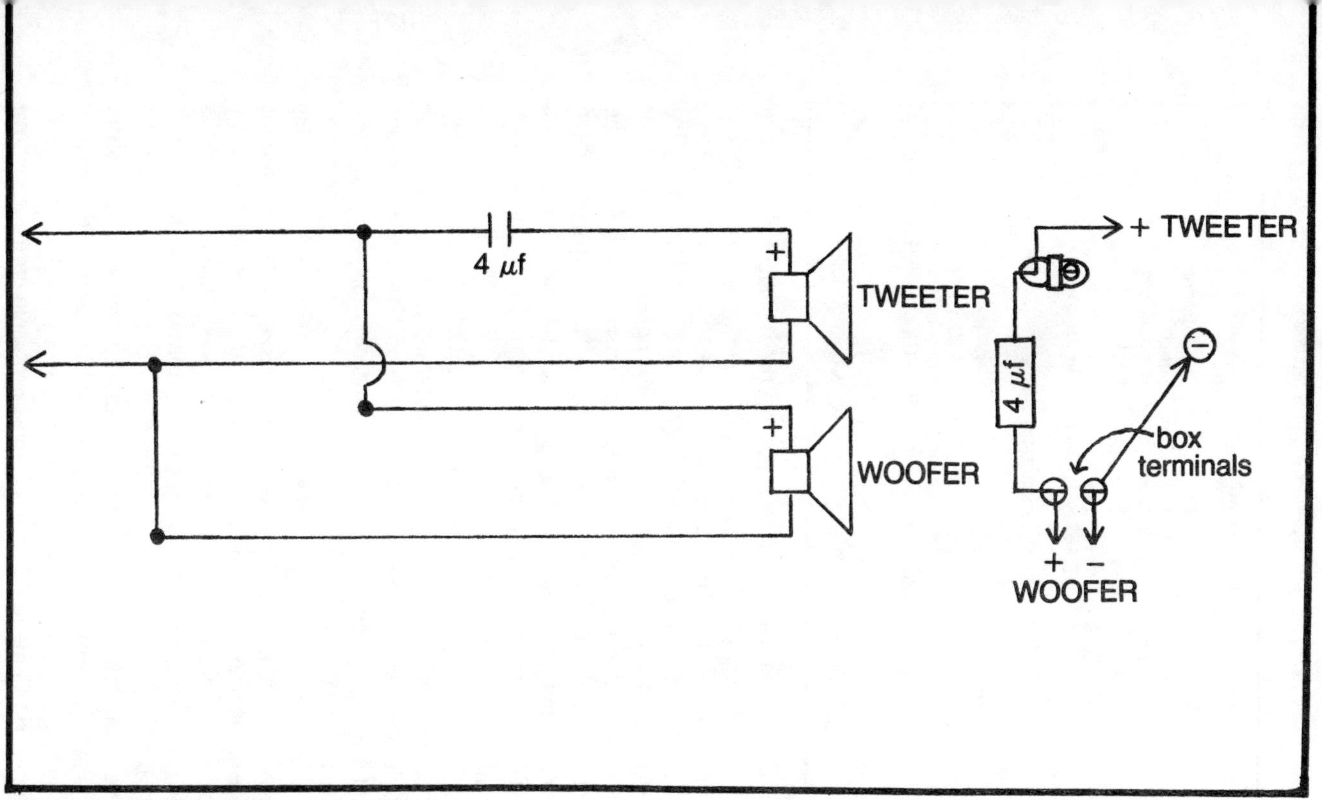

Fig. 9-8. Construction plans for Project 21.

Table 9-3. Materials list for Project 21.

	Enclosure Materials, ½″ Particle Board	
2	6-½″ × 7″	Top & bottom
2	7″ × 10-¼″	Sides
1	5-½″ × 10-¼″	Speaker board
1	6-½″ × 11-¼″	Back
⅜″ plywood		
1	5-7/16″ × 10-3/16″	Grille frame
	Speakers and Components	
1	4″ woofer, Norelco AD 4060/W4	
1	Dome tweeter, Norelco AD0163/T8	
1	4 μf non-polarized electrolytic capacitor.	
	Speakers available from Gladstone Electronics 1736 Avenue Rd., Toronto, ON M5M 3Y7	
60 grams of damping material lining walls of enclosure.		

Cut out the parts and set them together to check the fit. You may have to shave the edges of the speaker board to permit the other pieces to fit around it and be flush at the corners.

Mark a set of double lines, ½″ apart, on the inside and outside of each side, top and bottom, to indicate the position of the speaker board. Note that this set of lines should be set back ½″ from the front edge of each piece to make room for the front-mounted speakers and grille frame. These lines can be used to locate the speaker board on the inside of the panels and to place the nails on the outside.

Place the top and bottom pieces on a firm surface, and start #4-penny finishing nails at about 3″ intervals, placed between the double lines on the sides and top and bottom. Drive these nails in just far enough to penetrate the lower surface. Start other nails at a distance of ¼″ from each side edge of the top and the bottom.

Glue one side edge of the speaker board and the matching surface on a side. Nail the side to the speaker board. Glue the top edge of the side and speaker board and the matching surface on the top. Nail down the top. Continue this same procedure with the other side, and then do the same with the bottom (Fig. 9-9).

Drill two ¼″ holes through the center of the back, located the same distance from each other as the lugs on a screw terminal strip. Install a l-lug terminal inside the back an inch or so from the position of the holes. You can install this terminal with a small screw or with Heavy Grip Cement. Cut two 15″ pieces of lamp cord and remove

½″ of insulation from the leads. Pass both leads of one and one lead of the other through the holes in the back, from inside to outside. Observe polarity, and connect them to the lugs on the screw terminal strip. Connect a 4 μf capacitor to the terminal with one lead. Join the free end of the capacitor to the lug, and then connect the extra lead to the same lug. Solder the connections. Install the screw terminal strip on the outside of the back with glue and tacks.

Caulk the interior joints of the enclosure. Loosely fill the box with polyester fill damping material. Place the edges of the back on a firm surface and start #4-penny finishing nails around each edge. Thread the speaker leads through the damping material and out through the proper holes in the speaker board. Then glue and nail down the back to the enclosure.

FINISHING TOUCHES

Sand the exterior of the enclosure until it is smooth. Remove any loose dust or other foreign material before you try to stick on

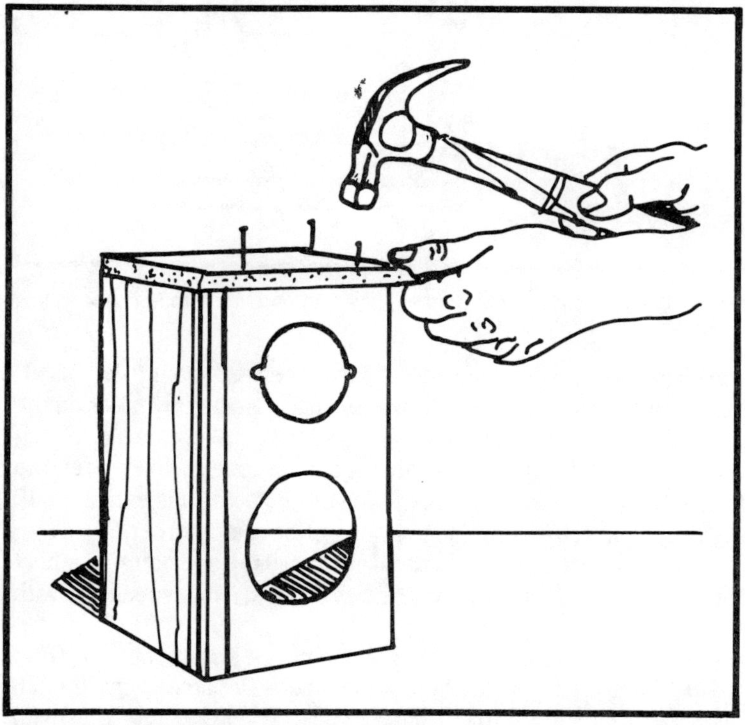

Fig. 9-9. Construction is by glue and nails.

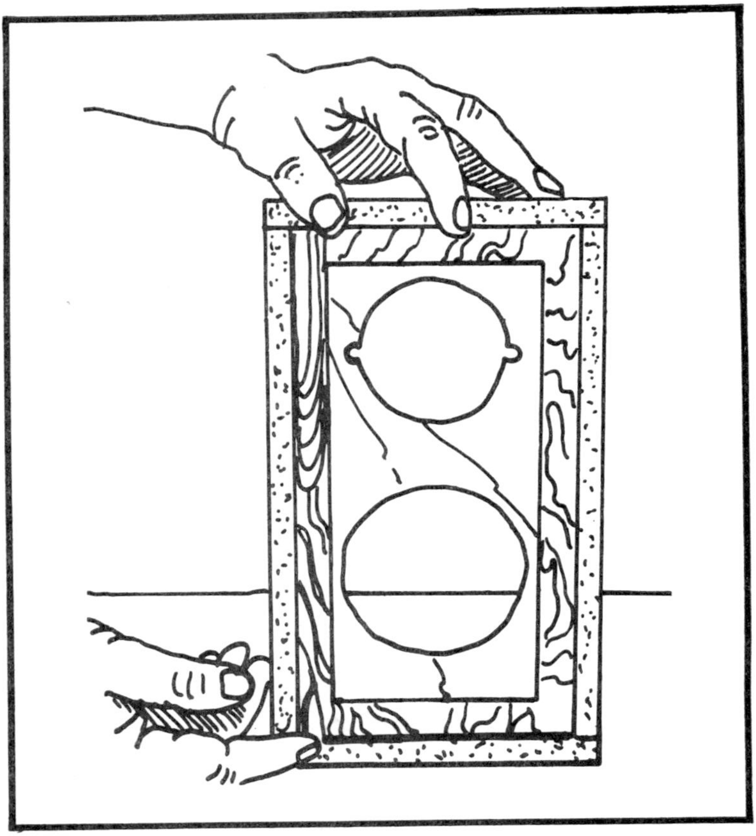

Fig. 9-10. Cut a grille frame from 3/8″ plywood and trim it to fit space in box loosely before cloth is installed.

the plastic veneer. Cut a piece of plastic veneer about 8-½″ × 36″ long. Start at the middle of the bottom and install the veneer on the box, allowing 1″ to overlap the front edge of the enclosure. Notch the corners and fold the protruding flap of veneer down over the front edges of the sides. Trim the top and bottom flaps to make a 45 degree angle to give the appearance of beveled corners.

Set the enclosure on its back and install the speakers. Connect the leads and solder them; then glue down the speakers with silicone rubber.

Cut a grille from 3/8″ plywood and check it for proper fit (Fig. 9-10). Paint it flat black. When dry, stretch a piece of grille cloth over the frame and staple it on the back. The grille should fit with enough friction to hold it in place.

This speaker, like the reflex model shown in Project 20, can deliver very clean wide range sound within its power range. For the closed box model, Norelco suggests an amplifier capable of 15 watts per channel or more. But you can use the speakers safely with more powerful amplifiers, particularly if you use them in a car or a very small room.

Choosing and Using

When you start out to build a speaker system, the obvious goal is to solder the last wire and put on the grille cloth. But don't stop there. The follow-up, getting the speakers set up and using them, is almost as important as wiring the speakers and building a good enclosure. But even before you start, what kind of speaker? If you have looked through the projects in this book and are trying to decide that, maybe a few suggestions will help.

CHOOSING A SPEAKER

To pick the right speaker you must consider the other components you will use, what kind of music you like, the number of speakers you intend to employ, and the kind of room you will put them in, but not necessarily in that order. To narrow the choice, determine the size of speaker that will be appropriate. Then consider other characteristics. Sometimes a speaker's impedance is pertinent. Or its efficiency. To make the right choice, examine each of these factors.

Size of Speaker and Room

The best advice on size is to keep things in proper scale. You obviously wouldn't want to try to fill a gymnasium with sound from a pair of mini-cubes. And two titans would be just as out of place in a van. Why try to defeat the laws of acoustics which tell us that large rooms are necessary for the lowest bass frequencies? Fortunately, there is a rule of thumb that can help you match the right size of speaker to the size of your room. If the proportions of speaker size and room size look right, they are probably right acoustically. To test the fit make up a mock speaker by cutting and taping together some pieces of shipping cartons. If the carton appears to be lost among the rest of the furniture, it is too small. If it overpowers everything around it, it's probably larger than necessary. But if you can arrange the carton so it fits nicely into your room and other furniture, chances are you've found the right size of speaker for satisfactory bass reproduction without wasting space.

Music Tastes

In addition to room size, you should consider what kind of music you like. Almost any size of good speaker can reproduce the tonal range and volume of a string quartet. A rock group or a symphony orchestra played back at high volume requires more output power than a small woofer can produce. If you like your music really loud, you probably need a speaker with a woofer that is 10″ in diameter or larger. The tonal range of rock music can be reproduced by a smaller woofer, but not at a high sound intensity level. So, in addition to choosing the right size of speaker for your room, you must pick a size that's right for you and your music. Do you want to hear the subtle differences in tone between similar instruments? Or, more than anything else, do you want power? The two characteristics rarely go together.

Power and Efficiency

If you have a receiver or amplifier with, say, a hundred or so watts per channel output, you don't have to be very much concerned with the efficiency of the speakers you build. If your amplifier has only marginal power, you will want to pick a speaker that has higher than average efficiency, such as Project 10 among the closed box speakers or Project 12 of the reflex speakers.

TWEETERS

If you have an amplifier with a relatively low power output and like your music loud, give some attention to the tweeter you choose. Small dome tweeters are especially prone to failure if overdriven. Tweeters are unlikely to be damaged in normal use if the crossover network is adequate. If you try to get 100 watts of power out of a 25-watt amplifier, something happens in the amplifier that can destroy your tweeter. That something is called clipping.

Suppose you have a tweeter that is recommended for duty at, say, 5000 Hz and up. And suppose your speaker system is rated 50 watts. If you have an amplifier that is rated at 25 watts, it can still blow your tweeter. Sound impossible? The fact is that most tweeters are rated for power levels far below that of the other drivers in a system. The normal power distribution against frequency is such that when 50 watts of power is supplied to a woofer, the power at 5000 Hz may be no more than a watt or so. But suppose your amplifier, when you overdrive it, starts to clip. When it clips, the sine wave changes to something resembling a square wave, with a production of odd harmonics. These odd harmonics occur at three times the fundamental frequency, five times the frequency and so on.

Now consider what happens if a 2500 Hz tone is clipped. This means that a 7500 Hz tone, a 12,500 Hz tone, and tones of still higher frequency are added. This adds, by distortion, power to the tweeter that should never have been in the signal. A more powerful amplifier would have delivered a pure 2500 Hz tone without the added burden on the tweeter.

If you have an amplifier with a total available power of, say, 50 watts or less and you want to shake the walls, you should choose a speaker with a rugged tweeter. Or substitute a tweeter that can take more power for the tweeter specified in a project. If you want your music that loud, you may not hear the subtle nuances of a small dome.

One possible substitution is to put a piezo-electric tweeter in the system. When using these tweeters at high power, it's a good idea to cut off low frequency energy to them. Speakerlab recommends a crossover network for piezo tweeters. Their network is diagrammed in Fig. 10-1. It crosses over at 9000 Hz, which means that you must have a mid-range driver that can perform well over an unusually wide range if this is your only tweeter.

Another possibility is to use a small cone speaker which has a relatively good high frequency performance. Some small cone speakers with whizzer cones serve this function satisfactorily. Their chief disadvantage is that they don't have as good dispersion as small tweeters.

You may have noticed that power ratings weren't given with comments on each speaker project. Such figures were omitted because they are often meaningless. A speaker system that can be safely driven by a 100 watt amplifier can be damaged by another 50 watt amplifier. You can get a rough guide by checking the advertised figures for the woofer used in each project.

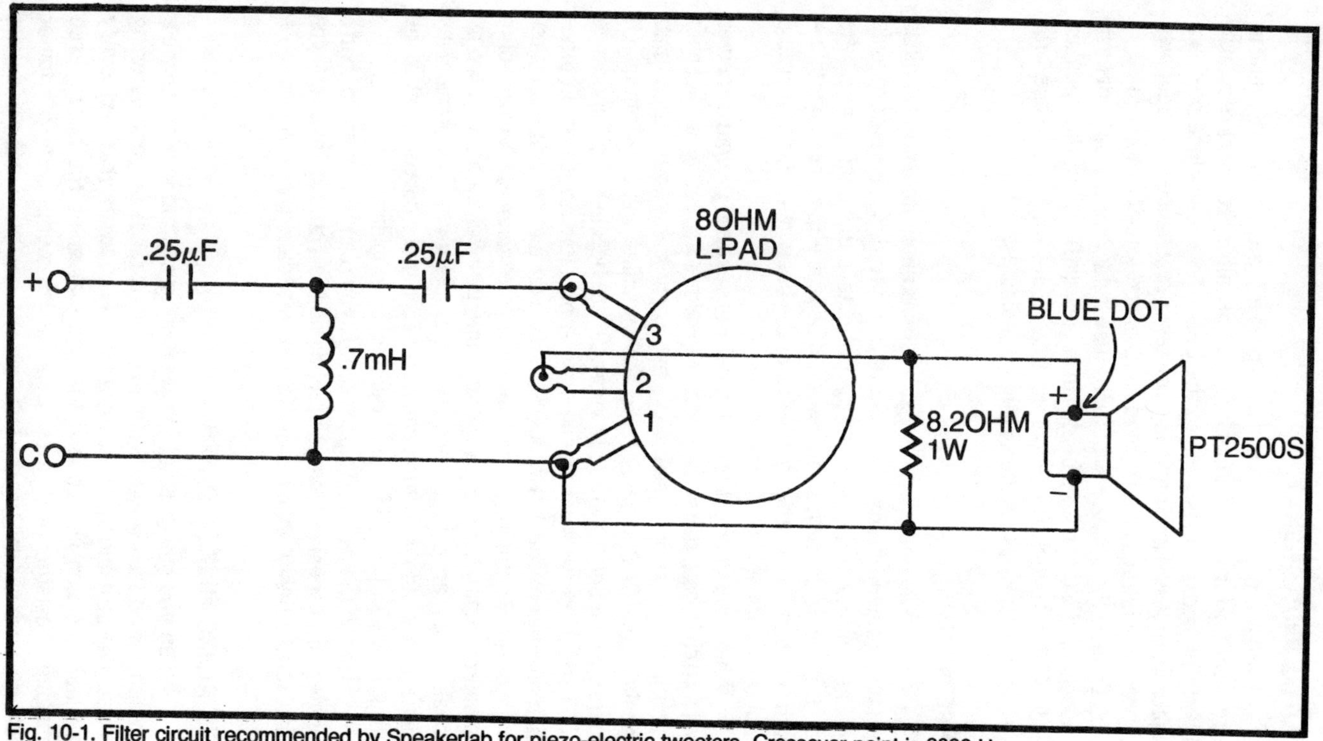

Fig. 10-1. Filter circuit recommended by Speakerlab for piezo-electric tweeters. Crossover point is 9000 Hz.

A LITTLE PROTECTION

If you are worried about your speaker, you can insert a fuse in series with it for protection The fuse may alter the tonal response slightly by adding resistance to the line. More resistance means a higher Q, or less damping on the woofer at resonance. The best way to predict whether this is good, bad or irrelevant is to try it.

To fuse the speakers, install a fuse holder on the back of each enclosure. Wire the fuse holder in series with one of the speaker leads. Table 10-1 shows some practical values for fuses to use with speakers of various power rating impedances.

IMPEDANCE

You can dispense with considerations about impedance unless you plan to use extension speakers or more than one pair of main speakers at one time. Some of the projects in this book have a note about impedance in the comments. Those are the ones that use 4 ohm woofers. The other speakers in the book can be considered 8 ohm speakers.

When you connect an extra set of speakers to your receiver and switch in both pairs, the second speakers appear in parallel with the main speakers. Assuming that each speaker is an 8 ohm speaker, the final impedance to the amplifier will be 4 ohms. Of course these figures are only approximate because the impedance of all speakers vary with frequency. A problem arises when more than one 4 ohm speaker or a 4 ohm speaker and an 8 ohm speaker are operated simultaneously in the same channel. Two 4 ohm speakers in parallel will give a net impedance of 2 ohms. A 4 ohm speaker plus an 8 ohm speaker in parallel gives 2.7 ohms. Any value below about 3 ohms can permit too much current flow in the output stage of a solid state receiver or amplifier.

Here is the rule on impedance. If you plan to use more than a single pair of speakers at one time, avoid the 4 ohm speakers. If you plan to use only a single pair, forget impedance.

INSTALLING YOUR SPEAKERS

When you buy or build a new speaker system you may have a preconceived idea about where to install them in your listening room. The first and most important rule is to *experiment*. Try every available location and choose the one that gives the best stereo imaging consistent with bass response. Sometimes even a minor change, such as twisting the cabinets to face a point in the middle of

Table 10-1. Fuse sizes for speakers of various power ratings.

Power rating of speaker	Speaker Impedance	
(Watts)	4 ohm	8 ohm
15	1-½ A	1 A
20	2 A	1-½ A
30	2 A	1-½ A
40	3 A	2 A
50	3 A	2 A
75	4 A	3 A

the opposite wall, can make a big difference in listening quality. Most modern speakers have good treble dispersion. When the speakers must be placed too far apart though, such tricks can improve performance. In a small room, where the speakers are too close to each other, twisting the cabinets in the opposite direction can sometimes help.

Try to install the speakers with the drivers, or at least the tweeters, near ear level. For compact speakers, this means some kind of support, such as a lamp table. There is usually no problem here except for the in-between size speakers, those too large for a table and too small for floor use. When such speakers are set on the floor against a wall much of the treble is lost and the bass can become too heavy. In such cases you can make an improvement by putting the speakers on a small stand. To find the optimum height, try stacking up some large books or other support until the sound is right. Then build a stand to provide the required elevation. It's a good idea to make the platform at the top of the stand slightly higher in front than in back to tip the tweeters upward. A tilt of about 5 to 7 degrees is about right for most rooms.

And so, having chosen your speakers well, install them properly, and put your mind at ease with some fuses; happy listening.

Appendix

Appendix

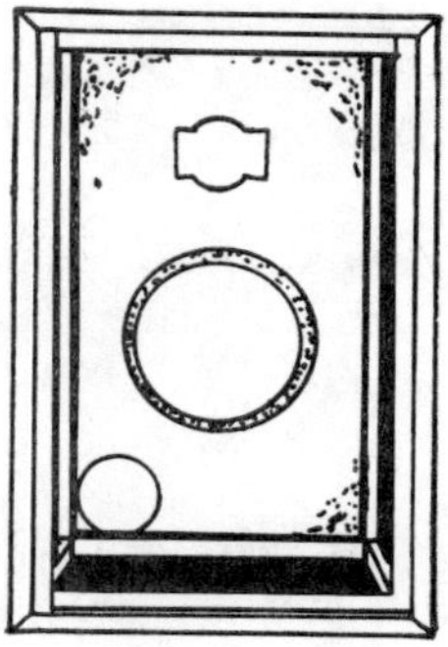

Here are sources of speakers and components. Some of these suppliers will send you a free catalog even though it carries a price tag of a dollar or so. If you write and ask for one, you will receive either a catalog or a notice of its price.

Acusta Craft, Inc.
P.O. Box 12030
Shawnee Mission, KS 66212
Offer a line of complete speaker system kits and drivers. Catalog.

Burstein-Applebee
3199 Mercier St.
Kansas City, MO 64111
Full line of electronics components, including magnet wire. Catalog.

EMS
P.O. Box 5301
Knoxville, TN 37918
Various drivers, including the Polydax line. Crossover networks and various values of coils and capacitors. Catalog.

Gladstone Electronics
1736 Avenue Rd.
Toronto, ON M5M 3Y7
Wide range of high fidelity speakers, including Altec, Celestion, KEF, Norelco, Peerless and others. Catalog.

McGee Radio Co.
1901 McGee St.
Kansas City, MO 64108
Extremely wide variety of drivers plus passive radiators. Some crossover networks and components. Catalog.

Olson Electronics *Line of house brand speakers. Catalog.*
260 S. Forge St.
Akron, OH 44327

Radio Shack *Line of house brand speakers and components. Get catalog from local store.*

Speakerkit, Ltd. *Line of speaker system kits plus various separate drivers including Electro-Voice, Phillips, and Polydax.*
P.O. Box 12, Route 4
Menomonie, WI 54751

Speakerlab, Department BRE *Wide variety of speaker system kits and separate drivers plus crossover components such as capacitors and coils. Also various low-priced booklets. Catalog.*
735 Northlake Way
Seattle, WA 98103

SRC Audio *A variety of name brand drivers such as Altec, Electro-Voice, CTS, Heppner, Panasonic Peerless, Phillips, Pioneer and Polydax. Catalog.*
3238 Towerwood Dr.
Dallas, TX 75324

Index

Index

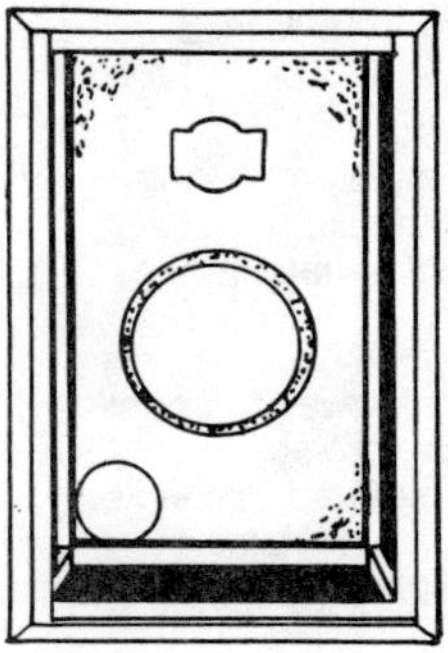